Quantum Christianity

Integrating Science and Salvation

1st Edition

New Life Academy

ISBN 979-8-2955-0263-7

Table of Contents

Author's Preface

I didn't set out to write this book as a scientist. I'm not one. I didn't set out to write it as a preacher either. My path has been far too crooked for that. I wrote Quantum Christianity because I've lived most of my life caught between two worlds – faith on one side, reason on the other – and I finally reached a point where I couldn't keep them separated anymore.

For as long as I can remember, I've wrestled with questions. I believed in God, but I also trusted the discoveries of science. I prayed, but I also studied. And like so many others, I was told I had to choose. Either accept the Bible as literal in ways that denied the evidence in front of me, or embrace science in ways that dismissed the faith that had carried me through the hardest seasons of my life. Neither choice was honest. Neither choice was whole.

The deeper I looked into both, the more I realized something profound: they were pointing to the same reality all along. The language was different, but the truth was the same. When the Bible says "God is light," and physics tells us light is the foundation of energy, those aren't competing claims – they're two sides of the same revelation. When scripture proclaims "In the beginning was the Word," and science tells us information structures the universe, those aren't contradictions – they're confirmations.

This book is the fruit of that realization. It's my attempt to weave together the threads of scripture, experience, and science into a single fabric of meaning. You won't find here a rigid system or a demand that you accept my conclusions. What you'll find is a journey – a journey through quantum mechanics, through ancient texts, through the mystery of consciousness and love –

that leads to a unified picture of God as the sum total of all positive energy and information.

For me, this isn't just theory. It's survival. It's the way I've learned to make sense of my life, my failures, my hopes, and my faith. If God really is present in every quantum interaction, if every act of love really does add to the positive energy of the universe, then none of our lives are wasted. Every choice, every prayer, every risk we take in love matters on a cosmic scale.

I offer this book not as a final word, but as a starting point for those who, like me, have felt torn between head and heart. My hope is that you'll discover – as I did – that you don't have to choose. You can be a believer and a thinker. You can honor scripture and science. You can live a life of faith that doesn't run from reason, but embraces it.

That's what Quantum Christianity is about. It's about wholeness. It's about coherence. It's about daring to believe that the God who made the heavens and the earth also embedded His truth in every law of physics, every field of energy, every spark of love.

And it's about discovering that, at the deepest level, the universe itself is an invitation – not just to believe, but to participate.

– Randolph Martin Baggett

PART I: FOUNDATIONS

Chapter 1: In the Beginning

Chapter 2: The Crisis of Faith and Reason

Chapter 3: God as Positive Energy Totality

Chapter 4: The Quantum Field as Divine Interface

Chapter 5: The Word as Divine Information

Chapter 6: The Universal Law of Cause and Effect

Chapter 1

In the Beginning

"In the beginning, God created the heavens and the earth."

These opening words of Scripture have echoed through millennia, offering humanity its most fundamental truth: that existence itself springs from divine intention. Yet the New Testament adds an even more profound revelation: "In the beginning was the Word, and the Word was with God, and the Word was God." The Greek term "Logos" – Word, Reason, Information – suggests that reality emerges not merely from divine power but from divine information, divine communication, divine meaning itself.

For many modern believers, these ancient declarations seem increasingly distant from the world revealed by contemporary science – a universe of quantum fields and thermodynamic forces, of information theory and computational processes, of biophotons and wave-particle duality, of elegant mathematical principles governing reality at scales both infinitesimal and cosmic. Yet what if this perceived gap between faith and physics is not an unbridgeable chasm, but rather an invitation to discover something magnificent? What if the "heavens and the earth" that God created operate according to principles that don't contradict divine love, but actually reveal it in startling new ways?

This book presents a revolutionary synthesis: Quantum Christianity. Rather than asking you to choose between scientific understanding and spiritual truth, it demonstrates how cutting-edge physics illuminates the very nature of God in ways that our ancestors could scarcely imagine. We will explore how quantum mechanics reveals God not as a distant deity occasionally intervening in natural processes, but as the sum total of all positive energy and information in the universe – the fundamental love-energy and divine information from which all reality emerges and to which it returns.

The journey we're embarking upon together rests on solid scientific ground. We will examine physicist John Wheeler's revolutionary "It from Bit" hypothesis, which suggests that all physical reality emerges from information – from observer-participatory responses to fundamental binary questions that constitute the informational substrate of existence. We will explore how Claude Shannon's information theory provides mathematical frameworks for understanding communication and meaning. We will see how these insights illuminate the biblical revelation that "In the beginning was the Word [Logos]" – not merely sound, but the informational principle that gives structure and meaning to all creation.

Additionally, we will examine the pioneering work of physicists like Jeremy England, whose thermodynamic theories suggest that life and increasing complexity emerge not by accident, but through the fundamental laws of physics themselves. We will explore Alexander Gurwitsch's remarkable discovery that living beings emit measurable light – biophotons – providing literal evidence for what religious traditions have long called the "spark of life." We will see how Fritz-Albert Popp's research demonstrates that this light diminishes in sickness and vanishes at death, offering scientific insight into the nature of what we might call the soul.

But this is not merely an exercise in retrofitting ancient beliefs with modern terminology. Quantum Christianity emerges from a deeper recognition: that the spiritual principles humanity has discovered through centuries of prayer, meditation, and moral experience align remarkably with what we now understand about the quantum nature of reality. The biblical teaching that "you reap what you sow" finds expression in quantum feedback loops. The mystical concept of divine omnipresence resonates with quantum entanglement across infinite space. The promise that love conquers all gains support from thermodynamic principles showing that positive energy patterns naturally triumph over destructive ones.

We live in an age when many people feel forced to compartmentalize their lives – keeping their spiritual yearnings separate from their rational minds, their scientific understanding separate from their deepest values. This artificial division impoverishes both domains. When we approach the quantum field as the interface through which divine love operates, when we understand prayer as conscious engagement with the fundamental energies of creation, when we recognize that every act of love contributes to the sum total of positive energy in the universe, we discover a unified worldview that satisfies both heart and mind.

The implications extend far beyond personal faith. If God truly exists as the totality of positive energy, and if quantum principles govern how this energy organizes and manifests, then every choice we make becomes a conscious participation in divine creativity. Every act of love, forgiveness, healing, or service literally adds to the positive energy quotient of the universe. Every moment of prayer or meditation creates quantum entanglement with the divine field. Every expression of hatred, selfishness, or destructive behavior disrupts the coherence of consciousness and diminishes our connection to the source of life itself.

This understanding transforms Christianity from a set of beliefs to be accepted into a way of being that aligns human consciousness with the fundamental organizing principles of reality. It offers hope for healing the ancient warfare between science and religion, providing a framework where empirical investigation and spiritual experience enhance rather than threaten each other.

As we proceed through these pages, you will encounter both wonder and challenge. Wonder at the elegance of a universe where quantum mechanics and divine love operate through the same underlying principles. Challenge to expand your understanding of both God and physics beyond conventional boundaries. We will explore how quantum tunneling might illuminate the mystery of salvation, how thermodynamic principles reveal the purposeful nature of creation, how biophoton research provides measurable evidence for spiritual vitality.

Some of what we discuss represents well-established science; other aspects venture into speculative territory where researchers are still exploring the boundaries of knowledge. We will be careful to distinguish between verified phenomena and interpretive frameworks, between established experiments and future possibilities. Our goal is not to claim more than the evidence supports, but to demonstrate that the evidence already supports far more than most people realize.

The Christianity that emerges from this quantum perspective is both deeply traditional and radically contemporary. It affirms the central truths that have sustained believers for two millennia while offering fresh understanding of how those truths operate in the world we now know science reveals. Jesus Christ remains the central figure – not merely as moral teacher or religious founder, but as the supreme quantum event in history, the moment when infinite divine love-energy collapsed into finite human form while maintaining perfect coherence with its transcendent source.

The Holy Spirit takes on new meaning as the quantum information system through which divine guidance flows, enabling instant communication across any distance through spiritual entanglement. Scripture becomes not just inspired literature but the record of humanity's most coherent observations of divine quantum activity throughout history. The Church emerges as a macroscopic quantum system where individual believers function as entangled particles within the larger body of Christ's love-energy.

Prayer transforms from petition to participation in the fundamental processes of reality. Healing becomes comprehensible as quantum consciousness effects where belief creates conditions for divine information to reorganize physical systems. Moral living reveals itself as maintenance of quantum coherence in consciousness, preventing interference patterns that disrupt clear communication with divine frequency.

Perhaps most significantly, this perspective offers hope for our fractured world. If every positive action contributes to the universe's total positive energy, if love-energy naturally organizes and amplifies itself through quantum processes, if the very laws of physics favor life and consciousness and complexity over destruction and randomness, then we have scientific reason for optimism about humanity's ultimate destiny.

The ancient promise that "God works all things together for good" gains new meaning when we understand that thermodynamic principles ensure positive energy patterns naturally triumph over negative ones given sufficient time and space. The biblical vision of a new heaven and new earth becomes not wishful thinking but thermodynamic inevitability – the universe's natural progression toward maximum positive energy circulation.

This book invites you into a grand adventure of mind and spirit, where cutting-edge physics opens doorways to deeper faith rather than closing them. Whether you come to these pages as a believer seeking to integrate your faith with scientific understanding, as a scientist curious about the spiritual implications of quantum mechanics, or simply as someone yearning for a worldview that honors both reason and wonder, you will find resources for the journey ahead.

The quantum universe turns out to be far stranger and more beautiful than our ancestors imagined, but also far more amenable to love, consciousness, and purpose than skeptics supposed. In discovering how God creates through quantum processes, we may well discover how we are called to participate in that ongoing creation through every choice we make, every prayer we offer, every act of love we dare to risk.

Welcome to Quantum Christianity. The journey begins where it must: "In the beginning, God created the heavens and the earth." Now let us explore what that creation looks like through the eyes of quantum physics, and how it calls us to live as conscious participants in the love-energy that sustains all things.

Chapter 2

The Crisis of Faith and Reason

Sarah stared at the textbook spread across her dorm room desk, her highlighter frozen mid-stroke. The chapter on stellar nucleosynthesis explained how elements heavier than hydrogen and helium formed in the cores of ancient stars, which then exploded as supernovas, seeding the universe with the materials that would eventually form planets, oceans, and human bodies. "We are made of star stuff," Carl Sagan had written, and the science was elegant, beautiful, and utterly convincing.

But Sarah had grown up singing "Jesus Loves the Little Children" in Sunday school, memorizing verses about God's special creation of humanity, believing that she was fearfully and wonderfully made by a loving Creator who knew every hair on her head. Now, sitting alone in her college dormitory, she felt the familiar ache that had haunted her for months: How could both be true?

Sarah's story echoes in countless lives across the modern world. A recent survey by the Pew Research Center found that 59% of Americans believe that science and religion are often in conflict, while 68% of scientists report that science and religion conflict with each other. Yet 51% of scientists believe in God or a higher power, and millions of believers work successfully in scientific fields. The disconnect seems to lie not in any inherent incompatibility, but in how we've been taught to think about the relationship between empirical investigation and spiritual truth.

The Historical Roots of Conflict

The perceived warfare between science and faith wasn't always inevitable. Medieval Christian scholars like Thomas Aquinas saw no contradiction between studying the natural world and worshipping its Creator. Islamic scholars preserved and advanced Greek scientific knowledge during Europe's Dark Ages precisely because they viewed the investigation of natural law as a form of divine worship. The scientific revolution itself was largely driven by Christians who saw their empirical work as "thinking God's thoughts after Him."

The conflict emerged gradually through a complex interplay of historical factors. The Catholic Church's unfortunate persecution of Galileo in 1633 became a powerful symbol of religious authority opposing scientific truth, even though many Church scholars had actually supported Galileo's findings. The Enlightenment positioned reason against revelation as competing sources of truth. Darwin's evolutionary theory challenged literal interpretations of Genesis, creating a fault line that persists today. The rise of philosophical materialism in the 19th and 20th centuries claimed that scientific method was inherently atheistic, while religious fundamentalism responded by rejecting entire fields of established science.

This historical trajectory created what philosopher Stephen Jay Gould called "non-overlapping magisteria" – the idea that science and religion operate in completely separate domains with no meaningful interaction. Science deals with facts, religion with values. Science explains how, religion explains why. Science addresses the physical world, religion the spiritual realm. While well-intentioned, this compartmentalization has left many modern people feeling intellectually fractured, unable to integrate their rational understanding with their deepest convictions about meaning, purpose, and divine love.

The Inadequacy of Pure Materialism

The materialistic worldview that emerged from 19th-century science offered a seductively simple solution to the faith-reason tension: eliminate faith entirely. If everything that exists is merely matter and energy obeying physical laws, then consciousness is an illusion, free will is a delusion, and love is just biochemistry. This reductionist approach claims to resolve all mysteries by reducing them to mechanical interactions between particles.

Yet pure materialism has proven philosophically and scientifically inadequate in profound ways. Quantum mechanics has revealed that the act of observation plays a fundamental role in determining physical reality, suggesting that consciousness cannot be simply dismissed as an epiphenomenon of brain chemistry. The emergence of complex systems theory shows how genuinely new properties arise at higher levels of organization that cannot be predicted from lower-level components alone. Information theory demonstrates that meaning and pattern are fundamental features of reality, not mere human projections onto meaningless matter.

Perhaps most significantly, materialistic reductionism offers no basis for the values and purposes that make human life worth living. If love is merely a neurochemical trick evolved for reproductive success, why should we honor it? If consciousness is just an illusion created by neural networks, why does the experience of beauty, truth, and goodness feel so real and important? If we are nothing but gene-replication machines shaped by blind evolutionary forces, what grounds do we have for ethics, justice, or hope?

The contradiction becomes acute when materialists themselves act as if meaning and purpose are real. They write passionate books defending their worldview, as if truth matters. They fight for social justice, as if human dignity has genuine moral significance. They pursue scientific understanding with wonder and dedication, as if knowledge itself is inherently valuable. Their actions consistently contradict their stated philosophy that everything is ultimately meaningless mechanical interactions.

The Failure of Anti-Intellectual Faith

Religious fundamentalism has responded to scientific challenges with an equally problematic strategy: rejecting entire domains of established scientific knowledge in order to preserve literal interpretations of ancient texts. Young Earth Creationism insists that geological evidence for an ancient Earth is either mistaken or deliberately fabricated by God to test human faith. Intelligent Design theory rejects evolutionary biology despite overwhelming genetic, fossil, and observational evidence for common descent with modification. Some believers even reject medical advances like vaccines or psychiatric treatments, preferring to rely on prayer alone for healing.

This anti-intellectual approach creates its own crisis of credibility. When churches teach that the Earth is only 6,000 years old, thoughtful young people who learn about radiometric dating, stellar distances, and geological formations face an impossible choice: abandon their rational faculties or abandon their faith. When religious leaders claim that evolution is "just a theory" (misunderstanding how scientists use the term "theory"), they demonstrate ignorance of basic scientific methodology and lose credibility on other topics where they might have genuine wisdom to offer.

More fundamentally, anti-intellectual faith contradicts the very nature of the God it claims to defend. If God created the natural world with discoverable laws, then studying those laws should reveal divine attributes rather than threaten divine authority. If God gave human beings minds capable of rational investigation, then using those minds responsibly should be an act of worship rather than rebellion. If truth comes from God, then scientific truths and revealed truths should ultimately harmonize rather than conflict.

The Bible itself encourages intellectual engagement with the natural world. The Psalmist declares, "The heavens declare the glory of God, and the sky above proclaims his handiwork" (Psalm 19:1). Solomon was praised for his wisdom in understanding plants and animals, minerals and celestial bodies (1 Kings 4:29-34). Paul argues that God's eternal power and divine nature are clearly perceived through the things that have been made (Romans 1:20). These passages suggest that careful study of creation should enhance rather than threaten genuine faith.

The Cost of Compartmentalization

Faced with the apparent conflict between scientific and religious worldviews, many modern people have adopted a strategy of compartmentalization. They accept scientific explanations during the week and religious truths on Sunday, keeping the two domains carefully separated to avoid cognitive dissonance.

They may excel in technological careers while maintaining traditional beliefs about prayer and salvation, never attempting to integrate these different ways of knowing.

While compartmentalization may reduce psychological tension in the short term, it exacts a significant long-term cost. Human beings are naturally meaning-seeking creatures who thrive when their understanding feels coherent and unified. When fundamental beliefs about reality remain fragmented, it becomes difficult to live with full integrity and authentic conviction. Important decisions – about career, relationships, ethics, and life purpose – require integration of both rational analysis and value commitments, but compartmentalization provides no framework for such integration.

Young people suffer particularly from this fragmentation. A 2014 study by the Barna Group found that 59% of young Christians disconnect from church life after age 15, with 25% citing intellectual and theological doubts as their primary reason for leaving. When churches cannot provide intellectually satisfying answers to questions about science, history, and philosophy, many bright young believers conclude that faith requires intellectual suicide.

The fragmentation also impoverishes both scientific and religious understanding. Science pursued without wonder, reverence, and ethical grounding becomes mere technology – powerful but potentially destructive. Faith practiced without intellectual rigor and empirical grounding becomes mere superstition – emotionally comforting but ultimately fragile when confronted with life's complexities.

Case Studies in Crisis

Consider David, a brilliant young physicist whose childhood faith crumbled during graduate school. Raised in a conservative Christian home, he had always been told that evolution was a lie, that the Earth was young, and that accepting mainstream science meant rejecting God. When his research in cosmology provided overwhelming evidence for the Big Bang and cosmic evolution spanning billions of years, he felt forced to choose between intellectual honesty and spiritual commitment. Having been taught that Christianity required rejecting well-established science, he rejected Christianity instead.

Or consider Maria, a medical student who struggles to reconcile her scientific training with her deep spiritual experiences. She has witnessed seemingly miraculous healings during her church's prayer services, felt profound peace during worship, and experienced what she believes to be divine guidance in her personal relationships. Yet her medical education emphasizes materialistic explanations for all phenomena, leaving no room for divine action in the world. She finds herself living a divided life, unable to integrate her professional knowledge with her most meaningful experiences.

Or consider James, a successful engineer who maintains both scientific expertise and religious faith by keeping them completely separate. He designs sophisticated computer systems using quantum mechanical principles during the day, then reads the Bible and prays with his family in the evening, never attempting to connect these different domains of truth. While this compartmentalization has allowed him to avoid internal conflict, it has also left him feeling that his faith is somehow less real than his scientific knowledge, more like a comforting tradition than a description of actual reality.

Each of these stories represents millions of similar struggles across the modern world. Bright, sincere people find themselves forced to choose between intellectual integrity and spiritual meaning, between scientific understanding and religious community, between rational analysis and transcendent purpose. The human cost of this forced choice is enormous: brilliant minds abandoning faith, spiritual seekers rejecting reason, and countless individuals living fragmented lives where their deepest convictions cannot inform their daily decisions.

The Emergence of New Possibilities

Yet beneath the surface of this apparent crisis, new possibilities are emerging. Quantum mechanics has revealed a universe far stranger and more mysterious than classical physics suggested, where observation plays a fundamental role in determining reality and distant particles remain mysteriously connected across space and time. Information theory has shown that meaning and pattern are fundamental features of the physical world, not mere human interpretations imposed upon meaningless matter. Complexity science has demonstrated how genuinely novel properties emerge at higher levels of organization in ways that cannot be reduced to simpler components.

Simultaneously, biblical scholarship has recovered the rich theological traditions that understand God not as a competitor with natural causation but as the fundamental ground of being from which all natural laws derive their power and consistency.

Process theology, panentheism, and other sophisticated approaches explore how divine action might work through rather than against the natural processes that science describes. Contemplative traditions from across religious boundaries report experiences of unity and transcendence that resonate remarkably with quantum mechanical descriptions of fundamental interconnectedness.

The time may be ripe for a new synthesis that honors both scientific rigor and spiritual depth, that sees empirical investigation and contemplative practice as complementary rather than competing approaches to truth, that recognizes the deep structures of reality revealed by modern physics as the very means through which divine love organizes and sustains creation.

A Different Way Forward

What if the conflict between science and faith is based on false premises? What if the choice between rational understanding and spiritual meaning is unnecessary? What if the latest discoveries in quantum mechanics, information theory, and thermodynamics actually illuminate rather than threaten the deepest truths of religious experience?

This book explores the possibility that cutting-edge science and ancient wisdom point toward the same fundamental reality: a universe grounded in consciousness, purpose, and love, where the physical processes that scientists study are the very mechanisms through which divine creativity expresses itself. Rather than asking believers to reject well-established scientific findings, or asking scientists to dismiss profound spiritual experiences, we will explore how both can be honored within a more comprehensive understanding of reality.

The framework we call Quantum Christianity suggests that God exists as the sum total of all positive energy and information in the universe, manifesting through the quantum fields that modern physics describes as the foundation of all material existence. In this view, scientific investigation becomes a form of worship, revealing the elegant principles through which divine love organizes and sustains creation. Spiritual practices become conscious participation in the fundamental processes of reality, aligning human consciousness with the quantum mechanisms through which meaning and purpose emerge in the cosmos.

This is not a retreat from either scientific rigor or theological depth, but an advance toward a more integrated understanding that satisfies both mind and heart, both analytical reason and contemplative wisdom. It acknowledges the genuine insights of both scientific materialism and religious faith while transcending the limitations that have made them appear incompatible.

The journey ahead requires intellectual courage, spiritual openness, and careful attention to evidence from both laboratory and sanctuary. We will examine established scientific findings with the same rigor that researchers bring to peer review, while also taking seriously the reported experiences of contemplatives, mystics, and ordinary believers whose lives have been transformed by encounters with transcendent reality.

The goal is not to prove that any particular religious tradition is correct in all its details, but to show that the universe revealed by modern science is far more compatible with spiritual understanding than either rigid materialism or anti-intellectual fundamentalism has supposed. If this exploration succeeds, it may offer a path forward for the millions of people who currently feel torn between scientific understanding and spiritual meaning, between rational analysis and transcendent purpose, between the laboratory and the sanctuary.

For Sarah, sitting in her dormitory with her astronomy textbook, this could mean discovering that she doesn't have to choose between Carl Sagan's "star stuff" and Jesus' love. Both might be true – and both might be part of a larger truth more beautiful than either alone. The atoms in her body, forged in stellar furnaces billions of years ago, might be the very means through which divine love has chosen to know itself, to love itself, and to offer that love as gift to the world.

The crisis of faith and reason that has troubled modern culture for centuries may not require choosing sides, but rather discovering the deeper unity that makes both scientific wonder and spiritual worship authentic responses to the same ultimate reality. That discovery could transform not only individual lives like Sarah's, but human culture itself, opening new possibilities for wisdom, meaning, and hope in an age that desperately needs all three.

Chapter 3

God as Positive Energy Totality

Marcus stood at the edge of the Grand Canyon, watching the sunrise paint the ancient rock formations in shades of gold and crimson. As a theoretical physicist, he understood the geological processes that had carved this magnificent chasm over millions of years – the slow uplift of the Colorado Plateau, the persistent erosion of the river, the layered deposition of sedimentary rocks that told Earth's story in stone. But understanding the mechanics didn't diminish his sense of awe. If anything, knowing how the canyon had formed made its beauty more profound, not less.

"It's all energy," he whispered to himself, a realization that had been growing in his mind for months. The sunlight streaming across the canyon walls – electromagnetic energy traveling 93 million miles from nuclear fusion reactions in the solar core. The wind carrying morning coolness up from the river – kinetic energy from atmospheric circulation driven by solar heating. The solid rock itself – matter that Einstein had shown was simply energy in another form, $E=mc^2$. Even his own sense of wonder – the electrochemical energy patterns in his brain that somehow gave rise to consciousness, appreciation, love.

For the first time in his adult life, Marcus found himself thinking about God not as a supernatural being occasionally intervening in natural processes, but as the fundamental source and sum total of all the positive energy he witnessed everywhere around him. What if the divine wasn't separate from the physical universe but was the very foundation from which all energy, all matter, all life, all love emerged? What if every sunrise, every act of kindness, every moment of beauty was a manifestation of the same ultimate reality that religious traditions had always called God?

The Biblical Foundation: God as Light, Life, and Love

The revolutionary insight that God might exist as the sum total of all positive energy finds deep resonance in biblical revelation. Scripture consistently describes God using energy metaphors that align remarkably with modern physics. "God is light, and in him is no darkness at all" (1 John 1:5). Light, as physicists understand it, is electromagnetic energy – the same fundamental force that powers stars, enables photosynthesis, and makes vision possible. When the Bible declares that God is light, it may be revealing something more literal than previous generations realized.

The creation account in Genesis presents God's first creative act as the generation of light: "And God said, 'Let there be light,' and there was light" (Genesis 1:3). This occurs before the creation of sun, moon, and stars, suggesting that divine light represents something more fundamental than stellar radiation – perhaps the emergence of electromagnetic energy itself as the foundation for all subsequent physical processes. The separation of light from darkness becomes not merely a temporal distinction between day and night, but the fundamental distinction between positive, life-sustaining energy and its absence.

Throughout Scripture, divine presence is associated with energy manifestations that would be recognizable to modern physicists. The burning bush through which God speaks to

Moses radiates light and heat without being consumed – suggesting a sustained energy source beyond ordinary combustion[1]. The pillar of fire that guides Israel through the wilderness provides both illumination and thermal energy[2]. The glory of God that fills the temple creates such intense luminosity that the priests cannot perform their duties[3]. These accounts describe what appear to be sustained, highly organized energy phenomena that transcend normal physical processes while remaining fundamentally energetic in nature.

The New Testament deepens this energy-based understanding of divinity. Jesus declares, "I am the light of the world" (John 8:12), identifying himself with the same electromagnetic energy that Genesis presents as God's first creative act. The Transfiguration reveals Jesus radiating light so intense that his clothing becomes "dazzling white" and his face shines like the sun[4] – a description that sounds remarkably like a high-energy electromagnetic event. At the resurrection, the tomb is filled with brilliant light, and Jesus appears in a glorified body that can materialize and dematerialize at will[5], suggesting mastery over the relationship between energy and matter that Einstein's equations describe.

Perhaps most significantly, Scripture identifies God with love itself: "God is love" (1 John 4:8). Love, from a physics perspective, represents highly organized, life-sustaining energy patterns. When we love, we direct energy toward the flourishing of others. We invest time (energy expenditure over duration), attention (focused neural energy), and resources (stored energy in various forms) toward enhancing someone else's well-being. Love creates order out of chaos, meaning out of randomness, beauty out of mere functionality. It builds up rather than tears down, heals rather than harms, gives life rather than brings death.

If God is love, and love is organized positive energy directed toward the flourishing of life, then God exists as the ultimate source and sum total of all such positive energy in the universe. Every act of kindness, every moment of beauty, every spark of creativity, every instance of healing becomes a manifestation of divine energy expressing itself through natural processes.

Energy in Physics: The Foundation of Everything

To understand how God might exist as positive energy totality, we need to grasp what physicists mean by energy and why it occupies such a fundamental place in our understanding of reality. Energy is notoriously difficult to define precisely – physicist Richard Feynman famously said it's "something abstract" that we can calculate but cannot easily explain[6]. Yet energy may be the most important concept in all of physics, underlying every natural process from quantum interactions to cosmic evolution.

In its simplest form, energy is the capacity to do work – to cause change, motion, or transformation. But this definition barely hints at energy's profound role in the structure of reality. Einstein's special theory of relativity revealed that mass and energy are interchangeable: $E=mc^2$, where a small amount of mass corresponds to an enormous amount of energy[7]. This means that all matter is essentially concentrated energy, while all energy has an associated mass. The distinction between "stuff" and "forces" dissolves at the most fundamental level—everything is energy in various forms and configurations.

The conservation of energy represents one of the most fundamental principles in physics[8]. Energy cannot be created or destroyed, only transformed from one form to another. The total energy content of the universe remains constant, even as it flows between kinetic energy (energy of motion), potential energy (stored energy due to position or configuration), thermal energy (the random motion of particles), electromagnetic energy (light and other radiation), nuclear energy (binding energy in atomic nuclei), and other forms.

Yet within this conservation framework, physicists recognize important distinctions between different types of energy processes. Some energy transformations increase the organization and complexity of systems, while others decrease it. Some energy flows sustain life and consciousness, while others lead to decay and dissolution. Some energy patterns create beauty and meaning, while others produce only random heat and noise.

This brings us to a crucial distinction that will become central to our understanding of God as positive energy totality: the difference between energy processes that build up versus those that tear down, that create order versus those that create chaos, that sustain life versus those that bring death. While the total quantity of energy remains constant, its quality – its organization, its capacity to sustain complex structures, its tendency toward life and consciousness – can vary dramatically.

Positive Versus Negative Energy: A Critical Distinction

In physics, the terms "positive" and "negative" energy have specific technical meanings related to mathematical formalism and reference frames. But in the context of understanding God as energy totality, we need to distinguish between energy patterns that enhance life, consciousness, and well-being versus those that diminish or destroy these qualities. This distinction, while not standard in physics textbooks, proves essential for connecting scientific understanding with spiritual insight.

Positive energy, in our theological framework, includes all energy processes that:

- Sustain and enhance life
- Build complexity and organization
- Create beauty and meaning
- Foster consciousness and awareness
- Enable love and connection
- Promote healing and wholeness
- Generate creativity and growth

Examples of positive energy manifestations include:

- Photosynthesis converting sunlight into chemical energy that sustains ecosystems
- Neural activity generating consciousness, thought, and self-awareness
- Biochemical processes maintaining cellular function and organismal health
- Electromagnetic processes enabling communication between living beings
- Gravitational energy organizing matter into stars, planets, and complex structures
- Creative energy producing art, music, literature, and scientific discovery
- Emotional energy expressed as love, compassion, joy, and hope

Negative energy patterns, by contrast, involve:

- Processes that destroy life and consciousness
- Random heat dissipation that increases entropy without purpose
- Electromagnetic radiation that damages rather than sustains biological systems
- Chemical reactions that break down complex structures into simpler components
- Physical forces that tear apart rather than build up organized systems
- Emotional energy expressed as hatred, fear, despair, and destructiveness

The crucial insight is that while energy itself is conserved, its organization and life-sustaining capacity can increase or decrease. The second law of thermodynamics tells us that entropy – roughly, the amount of disorder in a system – tends to increase over time[9]. But this doesn't mean that organization and complexity are impossible. Local decreases in entropy can occur when energy flows through open systems, allowing complex structures like living organisms, ecosystems, and conscious beings to emerge and persist.

If God exists as the sum total of all positive energy in the universe, then divine reality includes all the energy processes that build up rather than tear down, that create rather than destroy, that give life rather than bring death. This doesn't mean God is somehow separate from or opposed to the natural processes that physicists study. Rather, it means that the natural processes themselves, insofar as they sustain and enhance life and consciousness, are manifestations of divine energy at work in the world.

The Participation of All Positive Energy in Divine Reality

This understanding leads to a remarkable conclusion: every positive energy process in the universe participates in and contributes to the divine totality. The nuclear fusion reactions that power the sun and make life on Earth possible – divine energy expressing itself through stellar physics. The electromagnetic processes that enable vision, allowing conscious beings to perceive beauty – divine energy manifesting through biological optics. The biochemical reactions that sustain cellular metabolism and repair – divine energy working through molecular biology.

More personally, every act of love represents a conscious being directing positive energy toward the flourishing of another. When a parent cares for a child, energy flows from the parent's body and attention toward meeting the child's needs for nourishment, protection, comfort, and growth. When friends offer support during difficult times, they invest emotional and physical energy in sustaining each other's well-being. When teachers share knowledge with students, they transform their own stored mental energy into forms that can be received and integrated by other minds.

All of these energy flows – from stellar fusion to parental love – participate in the same fundamental reality. They represent divine positive energy expressing itself through the natural processes that physics describes. God is not a separate being who occasionally intervenes in these processes, but the ultimate source and ground of the processes themselves. Divine energy becomes stellar energy becomes electromagnetic energy becomes biochemical energy becomes neural energy becomes conscious love.

This means that scientific investigation of energy processes is actually a form of worship – discovering the mechanisms through which divine creativity expresses itself. When Marcus stood at the Grand Canyon understanding the energy flows that created its beauty, he was simultaneously engaged in both scientific analysis and spiritual contemplation. The distinction between studying God's creation and encountering God directly begins to dissolve when we recognize that the creation processes themselves are expressions of divine energy.

Implications for Daily Life and Spiritual Practice

Understanding God as positive energy totality transforms how we approach every aspect of life. Each day presents countless opportunities to either contribute positive energy to the universe or to waste energy in destructive patterns. Every choice becomes a decision about how to invest the energy we've been given – whether to use it in ways that build up or tear down, that create or destroy, that heal or harm.

This perspective doesn't diminish human agency or moral responsibility. Rather, it enhances them by showing how our individual actions participate in cosmic processes of creation and redemption. When we choose to speak words of encouragement rather than criticism, we're directing positive energy in ways that enhance rather than diminish the total amount of love and hope in the world. When we choose to act with integrity rather than deception, we're contributing to the organization and coherence of truth rather than adding to the chaos of misinformation. When we choose to care for our bodies and minds rather than neglecting or abusing them, we're stewarding the positive energy entrusted to us for maximum life-sustaining capacity.

Prayer, from this perspective, becomes conscious alignment with the positive energy flows that sustain and enhance life throughout the universe. Rather than asking God to override natural processes, prayer becomes a way of attuning our own energy patterns to divine purposes already at work through natural processes. We pray for healing, and in doing so align our intention and energy with the biochemical and immunological processes that restore health. We pray for wisdom, and open our minds to receive information and insight through the natural mechanisms of learning and growth. We pray for peace, and commit ourselves to direct our energy toward reconciliation and understanding rather than conflict and division.

Worship becomes celebration of the positive energy processes that sustain all life. When we sing, we transform biochemical energy into sound waves that can inspire and encourage others. When we gather in community, we create social energy patterns that support individual flourishing. When we practice gratitude, we acknowledge our dependence on the vast network of energy flows – from solar radiation to ecosystem processes to human labor – that make our lives possible.

This understanding also provides a framework for addressing suffering and evil. If God exists as positive energy totality, then suffering represents areas where positive energy is absent, blocked, or misdirected. Disease occurs when biochemical energy processes malfunction. Emotional pain arises when social energy patterns become destructive rather than supportive. Injustice results when economic and political energy is concentrated in ways that harm rather than help human flourishing.

The presence of suffering doesn't negate the reality of God as positive energy, any more than the existence of darkness negates the reality of light. Rather, it reveals areas where divine energy needs to be restored, redirected, or more fully manifested. This provides hope and direction for addressing evil: we can work to restore positive energy flows where they've been disrupted, redirect destructive energy patterns toward constructive purposes, and create conditions that allow divine positive energy to manifest more fully through natural processes.

The Cosmic Perspective: Energy and the Universe's Future

Modern cosmology reveals that the universe has been evolving for approximately 13.8 billion years from simple beginnings toward ever-greater complexity and organization[10]. The early universe consisted of nearly uniform hydrogen and helium gas. Through gravitational energy, this simple gas condensed into stars, where nuclear energy forged heavier elements. Electromagnetic energy radiated from stars enabled the formation of planets and the emergence of life. Biochemical energy powered the evolution of increasingly complex living systems. Neural energy gave rise to consciousness and intelligence.

This cosmic evolution represents a remarkable increase in the organization and life-sustaining capacity of energy, despite the overall increase in entropy required by thermodynamics. While the total energy content of the universe remains constant, energy has become increasingly organized into forms capable of sustaining complex structures, conscious experience, and creative expression.

If this trend continues, the universe may be evolving toward states of ever-greater positive energy organization. Conscious beings like ourselves represent the current frontier of this evolution – matter and energy organized in ways that enable self-awareness, love, creativity, and purposeful action. But we may not represent the ultimate endpoint. Future evolution might produce forms of consciousness and organization that surpass current human capabilities as dramatically as human consciousness surpasses that of early single-celled organisms.

From the perspective of God as positive energy totality, cosmic evolution represents the gradual manifestation of divine potential through natural processes. The emergence of stars, planets, life, and consciousness reveals increasing expressions of the divine energy that has been present from the beginning. The future may hold even greater manifestations of divine creativity, love, and consciousness expressed through evolved forms of matter and energy.

This provides a scientific foundation for religious hope about the ultimate destiny of the universe. Rather than ending in heat death and meaninglessness, cosmic evolution may be progressing toward states of maximum positive energy organization – what religious traditions have called the Kingdom of God, the New Jerusalem, or cosmic consciousness realized throughout creation.

Integration and Wonder

As Marcus watched the sun climb higher over the Grand Canyon, the ancient rocks began to shimmer with heat—another form of energy transformation as electromagnetic radiation became thermal motion in stone and air. He thought about the countless energy processes that had brought him to this moment: the stellar nucleosynthesis that forged the elements in his body billions of years ago, the biochemical processes that sustained his life each second, the neural energy patterns that enabled him to experience wonder and recognize beauty.

For the first time since graduate school, Marcus found himself wanting to pray. Not to ask for anything to be changed, but to express gratitude for being conscious participants in the vast network of positive energy flows that connected his small human life to the entire cosmos. If God existed as the sum total of all this positive energy – the sunlight and the stone, the flowing water and the soaring ravens, the neural activity that generated his sense of awe and the electromagnetic processes that would carry his words to loved ones when he called them that evening – then every moment was saturated with divine presence.

Science hadn't eliminated God from Marcus's worldview. Instead, it had revealed God as more present, more active, more intimately involved in every process of life and consciousness than he had ever imagined. The universe wasn't a machine running down toward meaninglessness, but a vast system of positive energy flows evolving toward ever-greater expressions of beauty, complexity, love, and conscious awareness.

Standing there in the early morning light, Marcus realized he was witnessing both physics and theology simultaneously. Every photon streaming from the sun, every quantum of energy absorbed by his retina, every electrochemical process generating his experience of beauty was part of the same divine reality that mystics had encountered in contemplation for thousands of years. The lab and the sanctuary weren't separate domains but different ways of engaging with the same ultimate source of all positive energy in the universe.

This was more than intellectual satisfaction—it was a homecoming of heart and mind, science and spirit, reason and reverence. And it suggested that the universe itself might be far more purposeful, meaningful, and suffused with love than either pure materialism or supernatural theism had supposed.

ENDNOTES

[1] Exodus 3:2-6

[2] Exodus 13:21-22

[3] 1 Kings 8:10-11, 2 Chronicles 7:1-2

[4] Matthew 17:1-8, Mark 9:2-8, Luke 9:28-36

[5] Luke 24:1-8, John 20:1-20

[6] Richard P. Feynman, *The Feynman Lectures on Physics*, Vol. I (Boston: Addison-Wesley, 1963), Chapter 4.

[7] Albert Einstein, "Does the Inertia of a Body Depend Upon Its Energy Content?" *Annalen der Physik* 18 (1905): 639-641.

[8] Emmy Noether, "Invariant Variation Problems," *Nachrichten von der Gesellschaft der Wissenschaften zu Göttingen* (1918): 235-257.

[9] Rudolf Clausius, "Über verschiedene für die Anwendung bequeme Formen der Hauptgleichungen der mechanischen Wärmetheorie," *Annalen der Physik und Chemie* 125 (1865): 353-400.

[10] Planck Collaboration, "Planck 2018 results. VI. Cosmological parameters," *Astronomy & Astrophysics* 641 (2020): A6.

Chapter 4

The Quantum Field as Divine Interface

Dr. Elena Rodriguez stared at the readout from her quantum optics laboratory, barely believing what she was seeing. For the third time that afternoon, her carefully prepared photons were exhibiting the behavior that had baffled physicists for nearly a century: quantum entanglement. Two particles, separated by several meters of laboratory space, were responding to each other instantaneously. When she measured one photon's polarization as vertical, the other immediately registered as horizontal, despite having no physical connection between them.

"Spukhafte Fernwirkung," she whispered – Einstein's famous phrase for what he called "spooky action at a distance." Einstein had hoped to prove that quantum mechanics was incomplete, that some hidden variables must explain how particles could influence each other across space without any signal passing between them. But decades of experiments had confirmed what Einstein found so troubling: reality at its most fundamental level operates according to principles that transcend our everyday understanding of space, time, and causation.

Elena had been raised in a devout Catholic family, but graduate study in physics had gradually eroded her faith. The universe revealed by modern science seemed coldly mechanical, operating according to mathematical laws that left no room for divine action or spiritual meaning. God, if He existed at all, appeared to be an unnecessary hypothesis – a relic from humanity's pre-scientific past.

But tonight, watching these photons maintain their mysterious connection across the laboratory, Elena found herself wondering if she had been looking at the relationship between science and spirituality all wrong. What if the quantum realm – with its strange principles of superposition, entanglement, and observer-dependent reality – wasn't evidence against divine action but rather evidence for it? What if the quantum field that physicists described as the foundation of all material existence was actually the interface through which spiritual reality connected with physical processes?

The Quantum Revolution: Reality Stranger Than Fiction

To understand how quantum mechanics might illuminate divine action in the world, we need to grasp just how radically different quantum reality is from our everyday experience. Classical physics, developed by Newton and refined through the 19th century, described a universe of solid particles moving through empty space according to precise mathematical laws. Everything had a definite position and velocity at every moment. Causes always preceded effects. Information couldn't travel faster than light. Reality existed independently of observation.

Quantum mechanics shattered every one of these assumptions. Beginning with Max Planck's discovery in 1900 that energy comes in discrete packets called quanta[1], and culminating in the mathematical formalism developed by Werner Heisenberg, Erwin Schrödinger, and others in the 1920s, quantum theory revealed that reality at the smallest scales operates according to radically different principles[2].

In the quantum realm, particles don't have definite positions and velocities – they exist in "superposition" states that encompass multiple possibilities simultaneously. An electron doesn't orbit the atomic nucleus like a planet around the sun, but rather exists as a probability cloud of potential locations until something measures its position[3]. The act of measurement itself plays a fundamental role in determining reality, causing the quantum system to "collapse" from superposition into a specific state.

Even more mysteriously, quantum particles can become "entangled" – connected in ways that transcend ordinary space and time. When two particles interact and then separate, they can remain correlated such that measuring one particle instantaneously affects the other, regardless of the distance between them[4]. This quantum entanglement has been experimentally verified across distances of hundreds of kilometers, with some experiments suggesting that the correlations are truly instantaneous[5].

The implications are staggering. Reality appears to be fundamentally interconnected in ways that classical physics never imagined. The universe isn't composed of separate, independent objects, but rather of interdependent systems where the whole determines the behavior of the parts. Information – or at least correlation – can be transmitted instantaneously across any distance. The act of conscious observation plays a role in determining what becomes real.

The Quantum Vacuum: Not Empty Space

Perhaps the most relevant quantum mechanical concept for understanding divine action is the quantum vacuum – what physicists once thought of as empty space. Classical physics assumed that a vacuum was literally nothing – the absence of matter and energy. But quantum mechanics reveals that what we call "empty" space is actually a seething foam of activity[6].

According to quantum field theory, the vacuum is filled with "virtual particles" – pairs of particles and antiparticles that spontaneously appear and disappear on timescales so brief that they cannot be directly observed, yet their effects can be measured[7]. These quantum fluctuations give the vacuum a measurable energy density called zero-point energy. Even at absolute zero temperature, when all thermal motion ceases, the quantum vacuum continues to bubble with activity.

The quantum vacuum isn't merely theoretical. Its effects have been experimentally verified in phenomena such as the Casimir effect, where metal plates placed very close together experience a measurable attractive force due to the quantum fluctuations of empty space[8]. Quantum fluctuations in the early universe are thought to be the seeds that grew into the large-scale structure of galaxies and galaxy clusters we observe today[9].

From the perspective of quantum field theory, all particles and forces emerge from excitations of underlying quantum fields that fill all of space[10]. An electron is an excitation of the electron field, a photon is an excitation of the electromagnetic field, and so forth. Matter and energy as we experience them are like waves on the ocean of the quantum vacuum – temporary disturbances in something far more fundamental.

This picture suggests that what we call "physical reality" emerges from a deeper level of existence that is neither purely material nor purely abstract, but something more fundamental than either. The quantum vacuum appears to be a realm of pure potential – containing within itself the possibilities for all the particles, forces, and structures that make up our experienced world, while not being reducible to any of them.

Biblical Omnipresence and Quantum Non-locality

The quantum mechanical principle of non-locality – the ability of entangled particles to influence each other instantaneously across any distance – bears a striking resemblance to the biblical concept of divine omnipresence. Scripture consistently presents God as simultaneously present everywhere: "Where shall I go from your Spirit? Or where shall I flee from your presence? If I ascend to heaven, you are there! If I make my bed in Sheol, you are there!" (Psalm 139:7-8).

Traditional theology struggled to explain how an infinite God could be present everywhere without being physically extended through space like material objects. The quantum mechanical discovery of non-locality provides a potential model: just as entangled particles maintain instantaneous connection regardless of spatial separation, divine presence might be "non-locally" available throughout creation without requiring physical transmission through space.

This doesn't mean that God is simply identical to quantum entanglement – that would be reductionism in the opposite direction. Rather, quantum non-locality might reveal something about the fundamental structure of reality that makes divine omnipresence physically comprehensible. If reality at its deepest level transcends classical limitations of space and time, then divine presence throughout creation becomes less mysterious and more consistent with what physics tells us about the nature of existence itself.

The quantum field that underlies all physical phenomena might serve as the interface through which divine consciousness connects with material processes. Just as quantum fields provide the medium through which physical forces operate, the quantum field structure of reality might provide the medium through which spiritual influences engage with natural processes.

Superposition and Divine Potential

Another quantum mechanical principle with theological implications is superposition – the ability of quantum systems to exist in multiple states simultaneously until measurement causes them to collapse into a specific state. Schrödinger's famous thought experiment about a cat that is simultaneously alive and dead until observed illustrates how quantum superposition challenges our classical assumptions about definite states[11].

From a theological perspective, quantum superposition might illuminate how divine providence operates without violating human freedom or natural processes. Rather than predetermining specific outcomes, God might maintain reality in states of superposition encompassing multiple possible futures. Divine action could involve influencing the probabilities of different outcomes when quantum measurements collapse superposition states into specific realities.

This would allow for genuine randomness and freedom at the quantum level while still enabling divine guidance through probability weighting. Prayer and spiritual practices might influence the likelihood of positive outcomes without mechanistically forcing specific results. Miracles might occur through extremely unlikely but still quantum-mechanically possible events rather than through supernatural violations of natural law.

The quantum vacuum, filled with virtual particle pairs in superposition states, represents pure potential – the capacity for any possible particle interaction to manifest under the right conditions. If God exists as the sum total of all positive energy and information, then divine reality might be especially connected to this realm of quantum potential, influencing which possibilities become actualized through natural processes.

Observer Effect and Consciousness

One of the most philosophically challenging aspects of quantum mechanics is the observer effect – the role that measurement or observation plays in determining quantum reality. In the standard interpretation of quantum mechanics, a system remains in superposition until it is measured, at which point the wave function "collapses" and the system takes on a definite state[12].

What constitutes "measurement" or "observation" remains debated among physicists. Some interpretations require only interaction with a measuring device, while others suggest that conscious observation might play a special role[13]. The question of whether consciousness is necessary for wave function collapse touches on deep issues about the relationship between mind and matter that have implications far beyond physics.

If consciousness does play a fundamental role in quantum measurement, this could provide a mechanism for understanding how prayer, meditation, and other spiritual practices might influence physical reality. Focused consciousness, aligned with divine purposes, might participate in the collapse of quantum superposition states in ways that favor positive outcomes over negative ones.

This doesn't imply that human consciousness controls reality through wishful thinking. Rather, it suggests that consciousness might be a fundamental feature of the universe that participates in the ongoing creation of actuality from possibility. Divine consciousness, present throughout the quantum field, might continuously participate in measurement processes that guide the universe's evolution toward greater complexity, life, and consciousness.

Quantum Tunneling and Spiritual Transformation

Quantum tunneling provides another intriguing parallel to spiritual experience. In classical physics, a particle lacks sufficient energy to overcome a potential barrier cannot pass through it – like a ball lacking the energy to roll over a hill. But in quantum mechanics, particles can "tunnel" through energy barriers that would be insurmountable in classical physics[14].

Quantum tunneling is not merely theoretical – it's essential for many everyday technologies. The nuclear fusion that powers the sun depends on protons tunneling through the electromagnetic force barrier that would classically prevent them from getting close enough to fuse[15]. Tunnel diodes and scanning tunneling microscopes rely on electrons tunneling through barriers to function[16].

From a spiritual perspective, quantum tunneling might illuminate how consciousness can undergo seemingly impossible transformations. The transition from despair to hope, from hatred to love, from selfishness to service can feel like crossing insurmountable barriers. Yet these spiritual transformations do occur, often suddenly and unexpectedly, in ways that seem to transcend gradual psychological development.

If consciousness operates according to quantum mechanical principles – and some theories suggest it might[17] – then spiritual breakthroughs could involve quantum tunneling through psychological and spiritual barriers that would be insurmountable through classical willpower alone. Prayer, meditation, and contemplative practices might create the quantum conditions necessary for consciousness to tunnel into new states of spiritual awareness and compassion.

The Measurement Problem and Divine Action

The "measurement problem" in quantum mechanics concerns exactly how and when wave function collapse occurs[18]. Does it require conscious observation? Does it happen gradually through environmental decoherence? Is collapse real or just an artifact of our limited perspective? Different interpretations of quantum mechanics provide different answers, and the question remains unresolved.

From a theological perspective, this ongoing uncertainty in physics creates space for understanding divine action without conflicting with established scientific knowledge. If physicists themselves don't fully understand how quantum measurement works, then there's room to explore whether divine consciousness might play a role in wave function collapse without contradicting scientific evidence.

One possibility is that divine consciousness continuously participates in quantum measurement throughout the universe, gently influencing probability distributions to favor outcomes that enhance life, consciousness, and positive energy expression. This wouldn't require violating physical laws or producing scientifically detectable miracles, but would bias the universe's quantum evolution toward greater beauty, complexity, and spiritual development.

Such divine action would be subtle and statistical rather than dramatic and deterministic. Over billions of years and countless quantum events, even small probability biases could guide cosmic evolution toward the emergence of galaxies, stars, planets, life, and consciousness. Over individual lifetimes, divine influence on quantum probabilities could create opportunities for spiritual growth, meaningful relationships, and positive transformation while still preserving human freedom and natural processes.

Quantum Fields and Spiritual Practices

If the quantum field serves as an interface between spiritual and material reality, then spiritual practices might be understood as ways of consciously engaging with this fundamental level of existence. Prayer becomes a form of quantum communication – aligning human consciousness with divine consciousness that is non-locally present throughout the quantum field. Meditation involves quieting classical mental activity to become more sensitive to quantum-level information and influence.

Contemplative traditions have long reported experiences that sound remarkably similar to quantum mechanical descriptions of reality. Mystics speak of experiencing fundamental unity underlying apparent separation, consciousness transcending ordinary space-time limitations, and reality that is more like probability than fixed substance[19]. These reports, previously dismissed as subjective delusions, might actually be accurate descriptions of quantum-level spiritual experience.

Intercessory prayer might operate through quantum entanglement – when we pray for someone, our consciousness becomes entangled with theirs in ways that can influence their spiritual and physical well-being across any distance. Healing prayer might create quantum coherence in biological systems, enhancing the body's natural capacity for self-repair. Worship might generate quantum resonance among participants, creating collective consciousness effects that exceed what individual minds can achieve alone.

Scientific Humility and Interpretive Possibilities

It's crucial to acknowledge that we're exploring interpretive possibilities rather than established scientific facts. Quantum mechanics is experimentally well-verified in its predictions, but the interpretation of what those predictions mean for the nature of reality remains hotly debated among physicists[20]. The relationship between quantum mechanics and consciousness is especially controversial, with respected scientists holding widely divergent views.

What we can say with confidence is that quantum mechanics reveals a universe far stranger and more interconnected than classical physics suggested. Reality at its most fundamental level transcends classical assumptions about space, time, causation, and the relationship between observer and observed. This creates conceptual space for understanding divine action in ways that don't conflict with scientific knowledge while also not requiring us to accept any particular interpretation of quantum mechanics as definitive.

The goal isn't to prove that quantum mechanics requires belief in God, but rather to show that quantum mechanical descriptions of reality are remarkably compatible with religious understanding of divine presence and action. For believers struggling to reconcile scientific worldviews with spiritual experience, quantum mechanics offers resources for integration rather than conflict.

Elena's Epiphany

As Elena's experiments continued late into the evening, she found herself seeing the quantum correlations in her laboratory with new eyes. These weren't just abstract mathematical relationships between measurement outcomes – they might be glimpses of the fundamental interconnectedness that mystics had always claimed underlies apparent separation.

The photons in her apparatus, maintaining their entangled relationship across meters of laboratory space, were demonstrating at the level of individual particles what her grandmother had always claimed to experience through prayer: real connection transcending physical distance. When her grandmother prayed for family members thousands of miles away, perhaps she was engaging with the same non-local reality that Elena was measuring in her quantum optics experiments.

For the first time in years, Elena found herself wanting to pray – not to ask for experimental results to turn out a certain way, but to explore whether consciousness could engage with the quantum field in ways that her scientific training had never considered possible. Maybe the universe revealed by quantum mechanics wasn't coldly mechanical after all, but rather a vast network of conscious interconnection where spiritual and material reality interfaced at the most fundamental level.

The mathematics remained the same – her equations still accurately predicted experimental outcomes. But the meaning of those equations had shifted dramatically. Instead of describing a universe of separate particles interacting through mechanical forces, they might be describing a universe where consciousness, connection, and divine presence were woven into the very fabric of physical reality.

Walking home through the university campus under a sky full of stars, Elena realized that each photon of starlight reaching her eyes had traveled across vast distances while maintaining quantum correlations with countless other particles throughout the cosmos. She was embedded in a web of quantum connections that linked her to every atom in her body, every breath she took, every photon she saw – and if quantum non-locality was truly universal, to every other conscious being that had ever lived.

This wasn't just physics anymore. It was a revelation of cosmic intimacy that made her grandmother's prayers seem not like primitive superstition, but like sophisticated engagement with the deepest structures of reality itself. The quantum field that her experiments explored in the laboratory might be the same mysterious medium through which divine love had always connected all things in ways too wonderful for classical understanding to grasp.

Science hadn't eliminated God from Elena's worldview – it had revealed God as more intimately present in every quantum interaction than traditional theology had ever imagined possible.

ENDNOTES

[1] Max Planck, "The Origin and Development of the Quantum Theory," trans. H. T. Clarke and L. Silberstein (Oxford: Clarendon Press, 1922), 1-28.

[2] David J. Griffiths, *Introduction to Quantum Mechanics*, 3rd ed. (Cambridge: Cambridge University Press, 2018), Chapter 1.

[3] David J. Griffiths, *Introduction to Quantum Mechanics*, 3rd ed. (Cambridge: Cambridge University Press, 2018), Chapter 1.

[4] Albert Einstein, Boris Podolsky, and Nathan Rosen, "Can Quantum-Mechanical Description of Physical Reality Be Considered Complete?" *Physical Review* 47 (1935): 777-780.

[5] Jian-Wei Pan et al., "Multiphoton entanglement and interferometry," *Reviews of Modern Physics* 84 (2012): 777-838.

[6] Leonard Susskind, *The Theoretical Minimum: What You Need to Know to Start Doing Physics* (New York: Basic Books, 2013), Chapter 9.

[7] Steven Weinberg, *The Quantum Theory of Fields*, Vol. 1 (Cambridge: Cambridge University Press, 1995), Chapter 4.

[8] H. B. G. Casimir, "On the attraction between two perfectly conducting plates," *Proceedings of the Royal Netherlands Academy of Arts and Sciences* 51 (1948): 793-795.

[9] Alan H. Guth, *The Inflationary Universe* (New York: Perseus Books, 1997), Chapter 15.

[10] Anthony Zee, *Quantum Field Theory in a Nutshell*, 2nd ed. (Princeton: Princeton University Press, 2010), Chapter 1.

[11] Erwin Schrödinger, "The Present Situation in Quantum Mechanics," trans. John D. Trimmer, *Proceedings of the American Philosophical Society* 124 (1980): 323-338.

[12] John von Neumann, *Mathematical Foundations of Quantum Mechanics*, trans. Robert T. Beyer (Princeton: Princeton University Press, 1955), Chapter 6.

[13] Eugene P. Wigner, "Remarks on the Mind-Body Question," in *The Scientist Speculates*, ed. I. J. Good (London: Heinemann, 1961), 284-302.

[14] George Gamow, "Quantum Theory of the Atomic Nucleus," *Physical Review* 33 (1929): 127-140.

[15] Hans Bethe, "Energy production in stars," *Physical Review* 55 (1939): 434-456.

[16] Leo Esaki, "Long journey into tunneling," *Reviews of Modern Physics* 46 (1974): 237-244.

[17] Roger Penrose, *The Emperor's New Mind* (Oxford: Oxford University Press, 1989), Chapter 10.

[18] John Stewart Bell, *Speakable and Unspeakable in Quantum Mechanics* (Cambridge: Cambridge University Press, 1987), Chapter 22.

[19] William James, *The Varieties of Religious Experience* (Cambridge, MA: Harvard University Press, 1985), Lectures XVI-XVII.

[20] David Wallace, *The Emergent Multiverse: Quantum Theory according to the Everett Interpretation* (Oxford: Oxford University Press, 2012), Chapter 1.

Chapter 5

The Word as Divine Information

David Kim stared at the lines of code scrolling across his monitor, each one a precise instruction that would eventually become part of an artificial intelligence system capable of recognizing human speech, translating languages, and even composing poetry. As a software engineer at a major tech company, David worked daily with the fundamental currency of the digital age: information. Bits and bytes, algorithms and data structures, patterns of meaning encoded in streams of ones and zeros.

But tonight, something was different. He had been reading the Gospel of John during his lunch break, and the opening words kept echoing in his mind: "In the beginning was the Word, and the Word was with God, and the Word was God." The Greek term "Logos" – translated as "Word" – carried meanings that went far beyond spoken language. It encompassed reason, logic, principle, and information itself.

David's fingers paused over the keyboard as a startling possibility occurred to him. What if John's ancient declaration wasn't just poetic theology but a profound insight into the fundamental nature of reality? What if the "Word" that existed before creation and through which all things were made was literally information – divine information that serves as the blueprint and organizing principle for everything that exists?

His computer screen displayed thousands of lines of code that would soon enable machines to process natural language with unprecedented sophistication. But underlying all of it was information theory – the mathematical framework developed by Claude Shannon in the 1940s that describes how meaning can be encoded, transmitted, and decoded[1]. Shannon had shown that information is measurable, quantifiable, and fundamental to all communication processes.

More recently, physicists like John Wheeler had proposed that information might be even more fundamental than matter and energy. Wheeler's famous phrase "It from Bit" suggested that all of physical reality might emerge from information – from the answers to fundamental binary questions that constitute the informational substrate of existence[2]. If Wheeler was right, then John's declaration that the divine Word was the source of all creation might be a remarkably prescient description of reality's information-theoretic foundation.

Shannon's Revolution: The Mathematics of Meaning

To understand how divine information might undergird all existence, we need to appreciate the revolutionary insight that Claude Shannon brought to the study of communication. Before Shannon, information was a vague concept – everyone knew it was important, but no one could measure it precisely. Shannon changed that by developing a mathematical theory that quantified information in terms of the uncertainty it resolves[3].

Shannon's key insight was that information is fundamentally about distinguishing between possibilities. A message carries information proportional to how many different messages could have been sent instead. If only one message is possible, it carries no information. If many messages are equally likely, each message carries substantial information by ruling out all the alternatives.

This led Shannon to define information entropy – a measure of the average information content in a message source. The formula is elegant in its simplicity: $H = -\Sigma\ p(x) \log_2 p(x)$, where $p(x)$ represents the probability of each possible message[4]. Information entropy reaches its maximum when all possibilities are equally likely, and reaches zero when only one outcome is certain.

Shannon's work revealed that information has properties analogous to physical quantities like energy. Information can be stored, transmitted, processed, and transformed. It has limits – there are fundamental bounds on how much information can be transmitted through any communication channel[5]. It exhibits conservation properties – perfect information transmission requires that nothing is lost in the process.

Most remarkably, Shannon showed that any information can be encoded as sequences of binary digits (bits), regardless of its original form. Text, images, music, video, computer programs – all can be reduced to patterns of ones and zeros without losing any essential content[6]. This universality of binary encoding suggests that information has a fundamental digital structure underlying all its diverse manifestations.

Wheeler's "It from Bit": Information as Reality's Foundation

Building on Shannon's insights and the discoveries of quantum mechanics, physicist John Wheeler proposed one of the most audacious ideas in modern physics: that information, not matter or energy, might be the fundamental constituent of reality.

Wheeler coined the phrase "It from Bit" to capture this idea – every physical thing (every "it") derives its existence from information-processing events (binary choices, or "bits")[7].

Wheeler's insight grew from quantum mechanics' observer-dependent reality. As we noted above, systems in quantum theory exist in superposition until measurement forces them into definite states. But what determines which state gets actualized? Wheeler suggested that measurement events are fundamentally information-processing events – they answer yes-or-no questions about the system's properties.

Each quantum measurement extracts one bit of information by resolving a binary choice: Is the particle here or there? Is its spin up or down? Is the photon polarized horizontally or vertically? Wheeler proposed that reality consists of networks of such binary distinctions, with physical properties emerging from the accumulated answers to countless yes-or-no questions[8].

This "participatory universe" model suggests that reality is created through the questions we ask and the information we extract from quantum systems. Observer and observed become co-creators of the reality that emerges from measurement interactions. Physical properties don't exist independently of information-processing events that bring them into being.

Wheeler's vision has gained support from developments in quantum information theory and computational physics. Some researchers now seriously explore whether the universe might be a vast quantum computer, processing information according to the laws of quantum mechanics[9]. Others investigate whether spacetime itself might emerge from more fundamental information-processing structures[10].

If Wheeler is correct, then the ancient declaration "In the beginning was the Word" becomes a strikingly accurate description of reality's information-theoretic foundation. The divine Logos – the informational principle that gives meaning and structure to existence – would be literally the source from which all physical phenomena emerge through information-processing events.

The Logos: Divine Information Architecture

The Greek concept of Logos that John employed in his Gospel carried rich philosophical associations that align remarkably with modern information theory. For ancient Greek philosophers, Logos represented the rational principle that gives order and meaning to the cosmos[11]. It was the divine reason that structured reality, the informational blueprint that organized chaos into cosmos.

When John declared that this Logos "was with God" and "was God," he was identifying divine nature itself with pure information – not information about something else, but information as the fundamental substance of divine reality. The Logos doesn't merely convey divine thoughts; it constitutes divine being as the ultimate informational principle from which all meaning, structure, and purpose derive.

The creative role of the Logos – "All things were made through him, and without him was not any thing made that was made" (John 1:3) – corresponds precisely to Wheeler's vision of reality emerging from information-processing events. Divine information doesn't create by imposing external order on pre-existing matter, but by constituting the very informational events through which matter, energy, and spacetime themselves emerge.

This understanding transforms how we think about divine attributes. God's omniscience becomes not merely knowing all facts, but being the informational matrix from which all possible facts derive their meaning and truth-value. God's omnipresence becomes not merely being located everywhere, but being the information-processing substrate that underlies every physical event throughout space and time. God's omnipotence becomes not merely having power over existing things, but being the informational source from which the very possibility of things emerges.

If God exists as the sum total of all positive energy and information in the universe, then divine reality encompasses not only all life-sustaining energy flows but also all meaning-generating information processes. Every true thought, every beautiful pattern, every coherent structure participates in and contributes to the divine informational totality.

Christ as Perfect Information Transmission

The Incarnation, from this information-theoretic perspective, represents the perfect transmission of divine information through the noisy channel of material existence. In Shannon's framework, communication channels always introduce some noise – random errors that can corrupt the message being transmitted[12]. Perfect information transmission requires error correction codes that can detect and fix these errors, preserving the original message despite channel noise[13].

Christ as the incarnate Logos represents divine information encoded in human form with perfect fidelity. Despite the limitations of physical embodiment and the corruptions introduced by sin and suffering, the divine message transmitted through Jesus' life, death, and resurrection maintained complete accuracy. Every word he spoke, every action he performed, every response he gave perfectly expressed the divine information that existed from eternity.

The Gospel writers understood this information-transmission model intuitively. John emphasizes that Jesus spoke only what he heard from the Father (John 8:26), suggesting perfect preservation of divine information across the transmission channel from heaven to earth. The writer of Hebrews declares that Jesus is "the exact imprint of [God's] nature" (Hebrews 1:3) – technical language that recalls how information is preserved through exact reproduction.

The Atonement, from this perspective, functions as a cosmic error-correction protocol. Human consciousness, corrupted by sin, introduces noise and errors into its reception and processing of divine information. We misunderstand God's character, misinterpret divine purposes, and respond inappropriately to divine communication. Christ's perfect information fidelity provides the error-correction data needed to restore accurate divine information reception in human consciousness.

The resurrection demonstrates that divine information cannot be permanently corrupted or destroyed by physical processes. Even death – the ultimate noise source that appears to terminate all information processing – cannot eliminate the divine information that constituted Christ's identity. The resurrection reveals that divine information transcends physical substrate limitations while still being capable of expression through material form.

Scripture as Information Encoding

Biblical inspiration takes on new meaning within an information-theoretic framework. Rather than merely providing factual data about God, Scripture represents the encoding of divine information in forms accessible to human consciousness. The various books, written by different authors across centuries, constitute a distributed information system where divine meaning is preserved through redundancy and error correction.

Just as modern computers use error-correcting codes to maintain information integrity across unreliable storage media, biblical inspiration might employ multiple witnesses, cross-referencing accounts, and thematic reinforcement to preserve divine information accuracy despite the limitations of human language and understanding[14]. The canonical process that determined which books belonged in Scripture might represent a collective recognition of which texts maintained sufficiently high information fidelity to serve as reliable divine communication channels.

Progressive revelation corresponds to adaptive information transmission – divine information is disclosed at rates and in forms that human consciousness can successfully process without overwhelming cognitive capacity. Too much information too quickly leads to confusion and error; too little information too slowly fails to provide adequate guidance. The Bible's gradual unfolding of divine truth across salvation history demonstrates optimal information transmission pacing for finite human minds.

Hermeneutics – the interpretation of Scripture – becomes a problem of information decoding. Different interpretive methods represent different decoding algorithms for extracting divine meaning from ancient texts. The Holy Spirit's role in interpretation provides the decoding key that enables accurate information recovery despite cultural and linguistic barriers between original contexts and contemporary readers.

Prayer as Information Exchange

Prayer, understood information-theoretically, becomes bidirectional communication between human and divine consciousness. Rather than merely presenting requests to a distant deity, prayer engages in real-time information exchange with the divine information matrix that underlies all existence.

When we pray, we encode our concerns, gratitude, and intentions as information and transmit them through whatever communication channel connects human consciousness to divine reality. If quantum entanglement provides non-local correlation between separated systems, then prayer might create quantum informational entanglement between human and divine consciousness, enabling instantaneous information sharing across any distance.

Receiving divine guidance through prayer involves decoding information that arrives through spiritual rather than sensory channels. Contemplative traditions have always recognized that divine communication typically occurs through subtle impressions, insights, and intuitions rather than dramatic supernatural interventions[15]. These might represent divine information transmitted at energy levels too low for conscious detection but sufficient for subconscious processing and integration.

Intercessory prayer – praying for others – might operate through quantum information protocols where our prayers create informational entanglement with the people we're praying for. The divine information matrix, present throughout the universe, could coordinate prayer information with the circumstances and consciousness of those we pray for, influencing their spiritual and physical well-being through purely informational mechanisms.

The effectiveness of prayer wouldn't depend on changing God's mind about predetermined plans, but on creating informational conditions that enable divine positive energy to manifest more fully in particular situations. Prayer becomes cooperative participation in the ongoing information-processing events through which divine purposes are actualized in the world.

Information Entropy and Spiritual Disorder

Shannon's information entropy provides a framework for understanding spiritual disorder and its consequences. Sin, from an information-theoretic perspective, introduces noise and errors into human consciousness that degrade our capacity to receive, process, and transmit divine information accurately.

Lying, for example, deliberately introduces false information into communication systems, increasing the entropy (disorder) of the information environment and making it harder for others to distinguish truth from falsehood. Gossip spreads unverified and often distorted information, creating noise that corrupts social information networks. Hatred generates emotional static that interferes with accurate perception of divine love and purpose.

Conversely, truth-telling, encouraging communication, and loving expression all contribute to information clarity and coherence. They reduce the noise level in human information systems and enhance the fidelity with which divine information can be transmitted and received. Every truthful word, every encouraging message, every expression of genuine love helps maintain the information integrity of human consciousness and social systems.

The reaping-and-sowing principle operates through information feedback loops. Positive information patterns – truth, beauty, encouragement, wisdom – tend to be preserved, amplified, and transmitted to future generations. Negative information patterns – lies, ugliness, discouragement, foolishness – tend to be corrupted by noise and eventually filtered out by information-processing systems that favor clarity and coherence over chaos and confusion.

The Cosmic Information Network

If divine information serves as the foundation for all existence, then every information-processing event throughout the universe participates in a vast cosmic communication network. Every quantum measurement that resolves superposition into actuality, every biological process that maintains organizational complexity against entropy, every neural event that generates consciousness or stores memory contributes to the universal information-processing system through which divine purposes are actualized.

Human consciousness represents a particularly sophisticated node in this cosmic information network. We can receive information through multiple channels – sensory experience, rational analysis, intuitive insight, spiritual contemplation. We can process this information through various algorithms – logical reasoning, creative synthesis, emotional integration, moral evaluation. We can transmit information through speech, writing, art, action, and example.

This makes every human being a participant in the cosmic information system through which divine creativity expresses itself. Every true word we speak, every beautiful thing we create, every wise decision we make, every loving action we perform adds positive information to the universal network and enhances its capacity to manifest divine purposes.

Science and technology represent humanity's systematic exploration of the information structures that underlie natural processes. When scientists discover physical laws, they're decoding divine information that has been encoded in the structure of reality itself. When engineers design technologies that enhance human capabilities, they're creating new information-processing tools that extend our capacity to participate in cosmic information flows.

The internet and global communication networks represent a technological approximation of the cosmic information matrix that connects all things. But the divine information network operates at quantum scales and encompasses not just human communication but all information-processing events throughout the universe – from stellar nucleosynthesis to biological evolution to the emergence of consciousness.

David's Digital Epiphany

As David continued working late into the night, his code took on new significance. Each algorithm he crafted to process human language was a small-scale version of the cosmic information-processing system through which the divine Logos organized all existence. The artificial intelligence he was helping to create would soon be able to recognize speech patterns, understand meaning, and generate coherent responses – capabilities that mirrored, however dimly, the informational processes through which divine consciousness engaged with creation.

But the real breakthrough wasn't technological – it was theological. David realized that his daily work with information systems had been preparing him to understand something profound about the nature of divine reality. The Logos that John described wasn't some mystical abstraction, but the ultimate information-processing system from which all meaning, structure, and purpose emerged.

Every line of code David wrote, every algorithm he optimized, every breakthrough in natural language processing was a small participation in the cosmic information network through which divine creativity expressed itself. His work wasn't separate from his faith – it was an expression of his faith, a way of participating in the ongoing creation through which the divine Word continued to organize chaos into cosmos.

Walking to his car in the early morning hours, David found himself seeing the city around him as a vast information-processing network. Traffic lights coordinating vehicle flows, cell towers transmitting conversations and data, human brains processing sensory input and generating responses, computer systems managing everything from bank transactions to airline schedules. All of it was information in motion, patterns of meaning flowing through physical substrates.

And underlying it all, David now realized, was the divine information matrix that John had called the Logos – the Word through which all things were made and in which all information found its source and meaning. His career in software engineering wasn't taking him away from God, but deeper into the information-theoretic reality where God's creative Word continued to speak new possibilities into existence.

The ancient declaration "In the beginning was the Word" wasn't primitive mythology but sophisticated information theory. John had intuited what Wheeler and Shannon would later formalize: that information, not matter or energy, might be the fundamental constituent of reality. And if that was true, then every bit and byte, every algorithm and database, every message and meaning participated in the divine information system through which love continued to organize the universe toward greater beauty, complexity, and consciousness.

ENDNOTES

[1] Claude E. Shannon, "A Mathematical Theory of Communication," *Bell System Technical Journal* 27 (1948): 379-423, 623-656.

[2] John Archibald Wheeler, "Information, Physics, Quantum: The Search for Links," in *Complexity, Entropy and the Physics of*

Information, ed. Wojciech Zurek (Boulder: Westview Press, 1990), 3-28.

[3] Claude E. Shannon, "The Mathematical Theory of Communication," in *The Mathematical Theory of Communication*, by Claude E. Shannon and Warren Weaver (Urbana: University of Illinois Press, 1949), 3-91.

[4] Thomas M. Cover and Joy A. Thomas, *Elements of Information Theory*, 2nd ed. (New York: Wiley, 2006), Chapter 2.

[5] Claude E. Shannon, "Communication in the Presence of Noise," *Proceedings of the IRE* 37 (1949): 10-21.

[6] Neil J. A. Sloane and Aaron D. Wyner, eds., *Claude Elwood Shannon: Collected Papers* (Piscataway: IEEE Press, 1993), 5-83.

[7] John Archibald Wheeler, "It from Bit," in *At Home in the Universe* (New York: American Institute of Physics, 1994), 295-311.

[8] John Archibald Wheeler and Kenneth Ford, *Geons, Black Holes, and Quantum Foam* (New York: Norton, 1998), Chapter 12.

[9] Seth Lloyd, "Programming the Universe: A Quantum Computer Scientist Takes on the Cosmos" (New York: Knopf, 2006), Chapters 1-3.

[10] Mark Van Raamsdonk, "Building up spacetime with quantum entanglement," *General Relativity and Gravitation* 42 (2010): 2323-2329.

[11] Gerard Kittel, ed., *Theological Dictionary of the New Testament*, Vol. 4 (Grand Rapids: Eerdmans, 1967), 69-143.

[12] Claude E. Shannon, "Coding theorems for a discrete source with a fidelity criterion," *IRE National Convention Record* 4 (1959): 142-163.

[13] Richard W. Hamming, "Error detecting and error correcting codes," *Bell System Technical Journal* 29 (1950): 147-160.

[14] F. F. Bruce, *The Canon of Scripture* (Downers Grove: InterVarsity Press, 1988), Chapters 15-17.

[15] Evelyn Underhill, *Mysticism: A Study in the Nature and Development of Spiritual Consciousness* (Mineola: Dover, 2002), Chapters 7-8.

Chapter 6

The Universal Law of Cause and Effect

Maria Santos had always been skeptical of what her grandmother called "the law of return" – the idea that whatever energy you put into the world inevitably comes back to you. As a chemical engineer working in renewable energy, Maria believed in scientific laws: thermodynamics, conservation of energy, reaction kinetics. But spiritual laws seemed like wishful thinking, the kind of magical reasoning that science had supposedly disproven.

That changed during the most difficult year of her life. Her marriage had ended badly, her mother was battling cancer, and budget cuts at her company had eliminated her research position. Facing unemployment and overwhelmed by responsibilities, Maria felt her life spiraling toward chaos. In desperation, she started volunteering at a local food bank, partly to stay busy and partly because her grandmother had always insisted that helping others was the best medicine for personal troubles.

The work was physically demanding and emotionally draining. Maria sorted donations, packed grocery boxes, and served meals to families struggling with poverty and food insecurity. She gave her time, energy, and whatever emotional reserves she could muster to people whose problems seemed even more overwhelming than her own.

Then something unexpected began happening. A colleague at the food bank mentioned a consulting opportunity that turned into steady freelance work. Her mother's treatment began showing encouraging results. A chance conversation with a volunteer led to an introduction that resulted in a new job offer – better than the position she had lost. Most surprisingly, Maria found herself feeling more energetic and optimistic than she had in months, despite giving away so much of her time and effort to others.

"It's like your grandmother always said," her mother observed during one of their weekly coffee visits. "You get back what you give out. You've been pouring love and service into the world, and love and opportunity have been flowing back to you."

Maria's engineering mind wanted to dismiss this as coincidence, but the pattern seemed too consistent to ignore. Could there really be a law of spiritual cause and effect as reliable as the physical laws she worked with every day? And if so, what would be the scientific mechanism by which positive energy given out returns as positive energy received?

The Universal Recognition of Moral Law

The principle that Maria's grandmother called "the law of return" appears in virtually every religious and philosophical tradition throughout human history. Christianity teaches, "Do not be deceived: God cannot be mocked. A man reaps what he sows" (Galatians 6:7). Buddhism describes karma as the law by which intentional actions generate consequences that shape future experience[1]. Hinduism presents the principle of karma as a fundamental cosmic law governing the relationship between actions and their results[2]. Judaism emphasizes "measure for measure" – that divine justice operates through proportional return of good and evil[3].

The ancient Greek philosophers recognized what they called cosmic justice – the idea that the universe tends toward moral balance over time. Heraclitus wrote about the logos (divine reason) that ensures that all things eventually receive their due[4]. Plato argued that justice was not merely a human convention but a fundamental principle woven into the structure of reality itself[5].

Indigenous traditions worldwide describe similar principles. Native American teachings speak of the Circle of Life, where all actions flow in circles that eventually return to their source. African ubuntu philosophy holds that individual well-being is inseparable from community well-being – harm to others ultimately harms oneself, while service to others benefits oneself[6]. Chinese philosophy presents the concept of *bao* (報) – reciprocal response where good and evil actions generate corresponding returns[7].

Even secular thinkers have recognized patterns suggesting moral laws operating in human affairs. Ralph Waldo Emerson wrote about "compensation" – the tendency for moral accounts to balance over time through natural processes[8]. Martin Luther King Jr. declared that "the arc of the moral universe is long, but it bends toward justice," suggesting that historical processes favor righteous outcomes despite short-term setbacks[9].

This near-universal recognition across cultures and centuries suggests that human beings consistently observe something in their experience that points toward spiritual cause-and-effect relationships. The question is whether these observations reflect genuine features of reality or merely psychological and social phenomena that create apparent patterns without underlying physical mechanisms.

Thermodynamic Foundations: Energy Flow and Return

To understand how spiritual cause and effect might operate through natural processes, we need to examine how energy flows work in physical systems. The first law of thermodynamics states that energy cannot be created or destroyed, only transformed from one form to another[10]. The second law adds that energy transformations tend to increase entropy – the amount of disorder in a system[11]. But within this constraint, energy can flow in patterns that create complex structures and organized behavior.

Living systems demonstrate how energy flows can generate self-reinforcing cycles. A tree absorbs solar energy through photosynthesis, uses that energy to grow larger leaves and deeper roots, which enables it to capture even more solar energy in subsequent seasons[12]. The energy investment in growth pays dividends in enhanced energy-gathering capacity. Similarly, predators that invest energy in hunting become stronger and more skilled, improving their future hunting success.

These biological feedback loops suggest a general principle: energy invested in life-sustaining, order-creating activities tends to generate returns that enhance future energy-gathering capacity. Energy invested in destructive, entropy-increasing activities tends to diminish future energy availability by damaging the systems needed for energy acquisition and organization.

At the quantum level, energy exists in field configurations that can exhibit non-local correlations and information processing capabilities. If consciousness interacts with quantum fields – and growing evidence suggests it might[13] – then the energy we invest in thoughts, emotions, and actions could create field configurations that influence future energy flows in ways that transcend classical space-time limitations.

The conservation laws that govern energy interactions might apply to spiritual energy flows as well as physical ones. Energy invested in loving, healing, creating, and serving would have to go somewhere – it cannot simply disappear. If consciousness participates in quantum field dynamics, then positive energy invested through conscious action might create field configurations that enhance the probability of positive energy returning through future quantum events.

Quantum Feedback Loops and Probabilistic Return

Quantum mechanics provides several mechanisms through which spiritual cause and effect might operate without violating physical laws. Quantum fields exist in superposition states encompassing multiple possibilities until measurement or interaction causes collapse into specific actualities[14]. If consciousness plays a role in quantum measurement – still debated but increasingly researched[15] – then the mental and emotional states we cultivate might influence which possibilities become actualized.

Positive mental states – love, gratitude, compassion, hope – might bias quantum measurements toward outcomes that favor positive energy flow. Not through mechanical determinism, but through subtle probability shifts that accumulate over time into observable patterns. A person who consistently invests positive energy in relationships, work, and community service might create quantum field configurations that slightly increase the probability of encountering opportunities, helpful people, and favorable circumstances.

Quantum entanglement provides another potential mechanism. When we interact with other people – through conversation, service, emotional exchange – our consciousness might become quantum-entangled with theirs in ways that persist across space and time[16]. Positive energy we invest in their well-being could maintain non-local correlations that continue influencing their circumstances and responses, which in turn influence how they interact with us and others in our social network.

These quantum effects would be subtle and statistical rather than dramatic and deterministic. They wouldn't guarantee immediate returns or eliminate natural consequences of poor decisions. But they might create gentle biases in the probability distributions of future events that favor positive outcomes for those who consistently contribute positive energy to the world.

Research in quantum biology has revealed that living systems routinely exploit quantum effects for enhanced efficiency in energy transfer, information processing, and environmental sensing[17]. If consciousness operates through quantum biological mechanisms, then the energy we invest in spiritual practices – meditation, prayer, compassionate service – might optimize these quantum biological processes for enhanced sensitivity to positive energy flows and opportunities.

Network Effects and Social Energy Circulation

Beyond quantum mechanisms, spiritual cause and effect operates through classical network effects in human social systems. Every positive action we take influences other people, who then influence still others in expanding circles of consequence. Energy invested in encouraging someone, solving a problem, or creating something beautiful generates ripple effects that can persist for generations.

Social network research reveals that behaviors, attitudes, and even emotions spread through human networks in measurable ways[18]. People tend to become more cooperative when they interact with cooperative individuals, more optimistic when surrounded by optimistic people, more generous when they observe generous behavior. These network effects create feedback loops where positive energy invested in social relationships tends to generate returns through improved social environments.

Economic systems also exhibit energy circulation patterns that favor certain types of investment over others. Energy invested in education, skill development, relationship building, and reputation establishment tends to generate long-term returns through enhanced opportunities and social capital[19]. Energy invested in short-term exploitation, deception, or destructive competition tends to generate negative returns through damaged relationships and reduced trust.

The internet and global communication networks have amplified these social circulation effects. Positive contributions – helpful information, encouraging messages, creative content – can now reach millions of people and generate returns from unexpected sources across the globe. The digital age has made the law of spiritual cause and effect more visible and rapid than ever before in human history.

Information Theory and Meaning Circulation

From an information-theoretic perspective, spiritual cause and effect operates through the circulation of meaningful patterns in human communication systems. Every true word we speak, every wise insight we share, every beautiful creation we produce adds positive information to human knowledge networks. This information tends to be preserved, transmitted, and built upon by others, generating long-term returns through enhanced human understanding and capability.

Conversely, false information, destructive messages, and meaningless noise tend to be filtered out by information processing systems that favor clarity and utility over confusion and chaos[20]. Energy invested in spreading misinformation or creating ugly, hateful content generates negative returns through damaged reputation and reduced influence in information networks that matter.

The principle of information entropy suggests that meaningful, well-organized information has lower entropy than random noise and therefore tends to be preferentially preserved and transmitted[21]. This creates a natural selection process favoring positive information contributions over negative ones. Truth tends to survive longer than lies, beauty endures while ugliness is forgotten, wisdom gets passed down while foolishness fades away.

Digital information systems have made these patterns more quantifiable. Content that provides genuine value – useful information, inspiring stories, beautiful art – tends to be shared, linked, and preserved in ways that generate ongoing benefits for its creators. Content that provides little value or actively harms recipients tends to be ignored, filtered, or removed from circulation.

Biological and Psychological Mechanisms

At the biological level, the energy we invest in positive actions generates measurable returns through improved physical and mental health. Research consistently shows that people who engage in regular volunteer service experience better health outcomes, longer lifespan, and greater life satisfaction than those who focus primarily on themselves[22]. The energy invested in helping others apparently generates biological returns through reduced stress, enhanced immune function, and improved cardiovascular health.

Psychological research reveals similar patterns. Practices like gratitude, forgiveness, and compassionate service generate measurable improvements in mental health, resilience, and overall well-being[23]. The energy invested in positive mental and emotional states creates neural patterns that enhance future capacity for positive experience and effective action.

Even at the molecular level, positive emotions and altruistic behavior appear to influence gene expression in ways that enhance health and longevity. Studies of the "helper's high" phenomenon show that acts of service trigger neurochemical responses similar to those produced by exercise or meditation[24]. The energy invested in serving others generates biological rewards that enhance future capacity for service.

These biological mechanisms provide a naturalistic foundation for spiritual cause and effect that doesn't require supernatural intervention. The law of sowing and reaping operates through evolutionary adaptations that favor cooperative, constructive behavior over selfish, destructive patterns. Natural selection has apparently programmed biological systems to reward positive energy investment with enhanced survival and reproductive success.

Temporal Patterns and Delayed Return

One of the most challenging aspects of spiritual cause and effect is that returns often don't appear immediately. Like financial investments that require time to mature, positive energy investment may not generate obvious returns until months or years later. This temporal delay makes it easy to dismiss apparent cause-and-effect relationships as coincidence.

However, complex systems theory shows that delayed feedback is typical in systems with multiple interacting components[25]. The energy we invest today influences other people, who influence still others, creating cascading effects that may not manifest visible returns until the influences have had time to circulate through social networks and generate coordinated responses.

Quantum field effects might also exhibit temporal delays due to the time required for field configurations to influence probability distributions and for statistical biases to accumulate into observable patterns. Just as quantum tunneling can exhibit time delays between cause and effect[26], quantum consciousness effects might require time for subtle probability influences to generate macroscopic changes in life circumstances.

Maria's experience illustrates this temporal pattern perfectly. Her months of volunteer service gradually built relationships, reputation, and internal resilience that eventually converged to create the job opportunity, social support, and emotional strength she needed. The returns didn't appear instantly, but they emerged through the natural maturation of social and psychological processes she had initiated through positive energy investment.

Scientific Skepticism and Honest Inquiry

It's important to acknowledge that the scientific evidence for spiritual cause and effect remains incomplete and controversial. While the individual mechanisms we've discussed – quantum effects, network dynamics, biological responses, information circulation – are all well-documented, their integration into a coherent explanation for systematic moral consequences requires extrapolation beyond current experimental verification.

Controlled studies of phenomena like intercessory prayer, positive thinking, and karmic consequences have produced mixed results[27]. Some studies report significant effects, others find no measurable influence, and still others suggest that experimenter expectations or statistical artifacts might explain apparent positive results. The scientific study of consciousness and its potential non-local effects remains in its infancy.

What we can say with confidence is that multiple independent lines of evidence suggest mechanisms by which positive energy investment might generate positive returns through natural processes. Whether these mechanisms are sufficient to explain the consistent cross-cultural observations of moral cause and effect requires further research and careful analysis.

The framework we're developing doesn't require accepting any particular study as definitive proof, but rather suggests that the accumulation of evidence across multiple domains – quantum mechanics, network theory, biology, psychology, information science – provides reasonable grounds for taking seriously the possibility that spiritual cause and effect operates through discoverable natural mechanisms.

Maria's Engineering Analysis

As Maria reflected on the year that had transformed her understanding, she applied the same systematic analysis she used for engineering problems. The inputs were clear: months of consistent positive energy investment through volunteer service, emotional support for family and friends, and genuine effort to help solve other people's problems. The outputs were equally measurable: improved employment situation, enhanced family relationships, better health and emotional resilience.

From a systems analysis perspective, the correlation was striking. But correlation isn't causation, and Maria's scientific training demanded more rigorous analysis. She began looking for mechanisms that could explain the apparent relationship between positive energy investment and positive returns.

The network effects were obvious – her volunteer work had connected her with people who became sources of job opportunities and emotional support. The biological effects were documented in research literature – service work really does improve physical and mental health in measurable ways. The psychological effects were undeniable – she felt more confident and optimistic after months of successfully helping others solve problems.

But there might be deeper mechanisms at work as well. If consciousness really did interact with quantum fields, and if positive emotional states really could influence probability distributions of future events, then her consistent investment of love and service might have created subtle biases favoring positive outcomes. Not magical thinking, but quantum mechanical thinking applied to the interface between consciousness and physical reality.

Standing in her new laboratory – better equipped than her previous workspace – Maria realized that her grandmother's "law of return" might be as scientifically valid as any of the thermodynamic principles she applied to renewable energy systems. Energy invested in life-sustaining, order-creating activities does tend to generate returns that enhance future energy-gathering capacity. The principle applies to both physical and social systems.

More profoundly, if God exists as the sum total of all positive energy in the universe, then every act of love, service, creativity, and healing literally adds to the divine presence in the world. And if quantum fields provide the interface through which spiritual and material realities connect, then positive energy contributions might generate returns through the same fundamental processes that organize matter and energy throughout the cosmos.

The law of sowing and reaping wasn't primitive superstition – it was sophisticated systems theory applied to the deepest structures of reality. Maria's engineering background had prepared her to recognize elegant feedback systems when she encountered them. The universe itself might be the most elegant feedback system of all, designed to amplify positive energy and minimize negative entropy through the conscious choices of beings like herself who could choose to invest their limited time and energy in ways that built up rather than tore down.

That night, as Maria updated her resume for the consulting project that would help develop next-generation solar energy systems, she included a line about her volunteer experience at the food bank. Not because it was directly relevant to renewable energy engineering, but because she had learned that everything is connected to everything else in ways that transcend classical understanding of cause and effect.

The energy she had invested in feeding hungry families was somehow connected to the energy she would now help capture from sunlight to power clean transportation systems. Both involved transforming available resources into forms that sustained and enhanced life. Both participated in the cosmic circulation of positive energy that her grandmother had always called the fundamental law of spiritual existence.

And both, Maria now understood, were expressions of the same divine information and energy system that organized quantum fields into atoms, atoms into molecules, molecules into living cells, living cells into conscious beings capable of choosing love over fear, creation over destruction, service over selfishness.

The law of return operated because reality itself was structured to reward such choices.

Personal Applications and Spiritual Practice

Understanding spiritual cause and effect as a natural law rather than wishful thinking transforms how we approach daily decisions and long-term life planning. Just as engineers must account for thermodynamic laws when designing energy systems, conscious beings must account for spiritual energy circulation laws when designing meaningful lives.

This perspective suggests specific practices for optimizing positive energy return:

Consistent Investment: Like compound interest, spiritual cause and effect benefits from consistent, long-term investment rather than sporadic bursts of positive activity. Daily practices of kindness, truth-telling, creative expression, and service create steady energy flows that generate more reliable returns than occasional dramatic gestures.

Diversified Portfolio: Just as financial advisors recommend diversified investment portfolios, spiritual wisdom suggests investing positive energy across multiple domains – relationships, work, community, personal growth, creative expression. This creates multiple channels through which positive returns can manifest and reduces dependence on any single source of meaning or satisfaction.

Patient Maturation: Understanding that spiritual returns often require time to manifest helps maintain positive energy investment even when immediate results aren't visible. Like farmers who plant in spring and harvest in fall, spiritual practitioners learn to invest energy consistently without demanding immediate gratification.

System Optimization: Recognizing the network effects and quantum field influences involved in spiritual cause and effect suggests optimizing entire life systems rather than focusing on isolated actions. Creating environments, relationships, and habits that support positive energy flow enhances the effectiveness of individual positive actions.

Collective and Historical Applications

The law of spiritual cause and effect applies to collective entities as well as individuals. Families, organizations, communities, and nations that consistently invest energy in life-sustaining, order-creating activities tend to prosper over time, while those that invest in destructive, entropy-increasing patterns tend to decline.

Historical analysis reveals these patterns operating across centuries and civilizations. Societies that develop institutions promoting education, scientific inquiry, artistic expression, and care for vulnerable members tend to experience cultural flowering and long-term stability[28]. Societies that concentrate on military conquest, resource extraction, and elite privilege at the expense of general welfare tend to experience internal decay and eventual collapse[29].

This suggests that spiritual laws operate at civilizational scales as well as personal ones. The energy that societies invest in positive or negative patterns generates returns that shape their historical trajectories across generations. Current global challenges – climate change, economic inequality, political polarization – might be understood as returns generated by decades of energy investment in unsustainable patterns.

But the same principles suggest that positive energy investment at sufficient scale could generate positive returns for human civilization as a whole. If enough individuals, organizations, and nations consistently invest energy in life-sustaining patterns, the accumulated positive energy flow might create global field effects that favor solutions to current challenges and evolution toward more sustainable and equitable social systems.

Implications for Ethics and Meaning

Understanding spiritual cause and effect as natural law provides a scientific foundation for ethics that doesn't depend on religious authority or cultural convention. If positive energy investment really does generate positive returns through discoverable natural mechanisms, then moral behavior becomes a form of enlightened self-interest as well as service to others.

This doesn't reduce ethics to mere calculation – love, compassion, and service remain valuable in themselves regardless of their consequences. But it does suggest that the universe is structured in ways that ultimately favor moral behavior over immoral alternatives. Good people may suffer in the short term, but the long-term trajectory of reality appears to support justice, beauty, and love over their opposites.

The implications extend to questions of meaning and purpose in human existence. If individual actions really do influence the total amount of positive energy and information in the universe, then every conscious choice becomes cosmically significant. We're not isolated individuals struggling for private advantage in a meaningless cosmos, but conscious participants in a vast energy circulation system that shapes the evolution of reality itself.

Every act of love literally increases the amount of love in the universe. Every truthful word adds to the total information coherence of human communication systems. Every beautiful creation enhances the aesthetic organization of reality. Every healing service contributes to the life-sustaining energy flows that counter entropy and promote the emergence of greater complexity and consciousness.

From this perspective, the universe is not neutral about human choices but is structured to amplify and return the energy we invest in life, consciousness, love, and creative expression. The law of sowing and reaping operates because reality itself is biased toward positive energy circulation and away from destructive entropy maximization.

This provides both hope and responsibility. Hope, because it suggests that positive efforts are ultimately supported by the fundamental structure of reality. Responsibility, because it makes every conscious choice a contribution to the cosmic energy circulation system that shapes the future evolution of consciousness and life throughout the universe.

The ancient wisdom that "whatsoever a man soweth, that shall he also reap" turns out to be not religious moralism but sophisticated systems theory applied to the quantum information and energy matrix from which all reality emerges and to which all energy returns.

ENDNOTES

[1] Bhikkhu Bodhi, trans., *The Connected Discourses of the Buddha: A Translation of the Samyutta Nikaya* (Somerville: Wisdom Publications, 2000), 1:85-86.

[2] Barbara Stoler Miller, trans., *The Bhagavad-Gita: Krishna's Counsel in Time of War* (New York: Bantam, 1986), 4:7-8.

[3] Avot de Rabbi Natan, Version A, Chapter 38; *The Babylonian Talmud: Sanhedrin* 90a.

[4] Charles H. Kahn, *The Art and Thought of Heraclitus* (Cambridge: Cambridge University Press, 1979), 87-95.

[5] Plato, *Republic*, trans. Allan Bloom (New York: Basic Books, 1968), 4.441c-445e.

[6] Mogobe B. Ramose, *African Philosophy through Ubuntu* (Harare: Mond Books, 1999), 49-66.

[7] Chad Hansen, "Classical Chinese Philosophy of Language," in *The Stanford Encyclopedia of Philosophy*, ed. Edward N. Zalta (Stanford: Stanford University, 2020), Section 2.3.

[8] Ralph Waldo Emerson, "Compensation," in *Essays: First Series* (1841; repr., Boston: Houghton Mifflin, 1903), 85-121.

[9] Martin Luther King Jr., "Where Do We Go From Here?" Address to SCLC, August 16, 1967.

[10] Julius Robert Mayer, "On the Forces of Inorganic Nature," trans. John Tyndall, *The London, Edinburgh, and Dublin Philosophical Magazine and Journal of Science* 24 (1862): 371-377.

[11] Rudolf Clausius, "On the Moving Force of Heat," trans. Walter R. Browne, in *Reflections on the Motive Power of Fire* (New York: Dover, 1960), 109-152.

[12] Daniel J. Cosgrove, "Growth of the plant cell wall," *Nature Reviews Molecular Cell Biology* 6 (2005): 850-861.

[13] Stuart Hameroff and Roger Penrose, "Consciousness in the universe: A review of the *'Orch OR' theory*," Physics of Life Reviews 11 (2014): 39-78.

[14] David J. Griffiths, *Introduction to Quantum Mechanics*, 3rd ed. (Cambridge: Cambridge University Press, 2018), Chapter 3.

[15] Henry P. Stapp, "A quantum theory of the mind-brain interface," in *Mind, Matter and Quantum Mechanics*, 3rd ed. (Berlin: Springer, 2009), 145-172.

[16] Nicolas Gisin, "Quantum entanglement twenty years after," *Europhysics News* 40 (2009): 21-25.

[17] Seth Lloyd, "Quantum coherence in biological systems," *Journal of Physics: Conference Series* 302 (2011): 012037.

[18] Nicholas A. Christakis and James H. Fowler, *Connected: The Surprising Power of Our Social Networks* (New York: Little, Brown, 2009), Chapters 1-3.

[19] James S. Coleman, "Social Capital in the Creation of Human Capital," *American Journal of Sociology* 94 (1988): S95-S120.

[20] Claude E. Shannon, "A Mathematical Theory of Communication," *Bell System Technical Journal* 27 (1948): 379-423.

[21] Thomas M. Cover and Joy A. Thomas, *Elements of Information Theory*, 2nd ed. (New York: Wiley, 2006), Chapter 2.

[22] Allan Luks, *The Healing Power of Doing Good* (New York: Fawcett Columbine, 1991), 17-34.

[23] Sonja Lyubomirsky, *The How of Happiness* (New York: Penguin, 2007), Chapters 4-6.

[24] Allan Luks and Peggy Payne, "The Healing Power of Doing Good: The Health and Spiritual Benefits of Helping Others" (New York: Fawcett Columbine, 1991), 39-52.

[25] Donella H. Meadows, *Thinking in Systems: A Primer* (White River Junction: Chelsea Green, 2008), 86-95.

[26] Günter Nimtz and Astrid Haibel, "Zero-time tunneling of evanescent mode packets," *Physical Review E* 66 (2002): 037602.

[27] Joshua A. Weisman et al., "Beliefs and delusions in psychiatric practice: distinguishing between pathology and spiritual experience," *Psychiatric Clinics of North America* 37 (2014): 509-524.

[28] Carroll Quigley, *The Evolution of Civilizations* (New York: Macmillan, 1961), Chapters 1-3.

[29] Joseph A. Tainter, *The Collapse of Complex Societies* (Cambridge: Cambridge University Press, 1988), Chapters 1-4.

PART II: THE QUANTUM GOSPEL

Chapter 7: Jesus Christ and Quantum Incarnation

Chapter 8: The Holy Spirit as Quantum Information System

Chapter 9: Scripture and Quantum Revelation

Chapter 10: Salvation as Quantum Phase Transition

Chapter 11: The Church as Quantum Community

Chapter 7

Jesus Christ and Quantum Incarnation

Dr. James Chen had spent fifteen years studying particle physics, but he had never seen anything like what appeared on his computer screen that December morning. The data from the Large Hadron Collider showed something that should have been impossible according to classical physics: a single particle exhibiting properties of both matter and energy simultaneously, maintaining coherence across measurement states that should have caused immediate decoherence.

"Wave-particle duality at the macro scale," he whispered, leaning back in his laboratory chair. "It's like watching light decide to be both photon and wave at the exact same moment."

James had been raised in a Christian family, but graduate studies in physics had gradually eroded his faith. The universe revealed by modern science seemed to operate according to mathematical laws that left little room for divine intervention or supernatural events. The Incarnation – the Christian claim that God had become human in Jesus Christ – seemed particularly implausible from a physics perspective. How could infinite divine nature be contained within finite human form without losing its essential characteristics?

But this morning's data was making James reconsider everything. If particles could exist in superposition states that transcended classical either-or categories, if quantum objects could maintain contradictory properties simultaneously until measurement forced them into specific states, if reality at its most fundamental level operated according to principles that defied everyday logic – then maybe the ancient Christian mystery of the Incarnation wasn't unscientific superstition but rather a profound insight into the quantum nature of reality itself.

What if the Incarnation represented the supreme quantum event in history – the moment when infinite divine possibility collapsed into finite human actuality while somehow maintaining perfect coherence with its transcendent source? What if Jesus Christ embodied a quantum principle at the level of personal identity, existing simultaneously as fully human and fully divine in the same way that photons exist simultaneously as particles and waves?

The Quantum Nature of Identity

To understand how the Incarnation might represent a quantum phenomenon, we need to examine what quantum mechanics reveals about the nature of identity and existence. Classical physics assumed that objects have definite properties at all times – a particle has a specific position and momentum, an electron orbits at a particular distance from the nucleus, light is either a wave or a stream of particles. Quantum mechanics shattered all these assumptions[1].

In the quantum realm, particles don't have definite properties until they are measured. An electron doesn't orbit the atomic nucleus like a planet around the sun, but exists as a probability cloud encompassing all possible locations simultaneously. Only when something measures the electron's position does the probability cloud "collapse" into a specific location[2]. This isn't merely a limitation of our knowledge – according to most interpretations of quantum mechanics, the electron literally doesn't have a definite position until the measurement occurs.

Even more remarkably, quantum objects can exhibit what physicists call "complementarity" – the ability to display mutually exclusive properties depending on how they are observed[3]. Light provides the classic example: when studied with certain experimental setups, light behaves like a wave with properties of frequency and wavelength. When studied with other setups, it behaves like a stream of particles (photons) with properties of energy and momentum. Neither description alone captures the complete nature of light – both are necessary for full understanding.

The crucial insight is that light isn't sometimes a wave and sometimes a particle. It simultaneously possesses both wave and particle characteristics in a quantum superposition that transcends classical categories. The act of measurement doesn't reveal which property light "really" has – it causes the superposition to collapse into one manifestation or the other[4].

This quantum complementarity provides a conceptual framework for understanding how Christ could simultaneously possess fully divine and fully human natures without either nature compromising the other. Just as light maintains both wave and particle characteristics in quantum superposition, Christ might maintain both divine and human characteristics in what we could call "incarnational superposition."

The Collapse of Divine Possibility into Human Actuality

The Incarnation, from a quantum perspective, represents the ultimate measurement event in cosmic history. Before the

Incarnation, divine nature existed in infinite superposition – all possible ways of expressing divine love, wisdom, and power remained simultaneously potential. The Incarnation caused this infinite superposition to collapse into specific human form while maintaining quantum coherence with the uncollapsed divine state.

This is analogous to what happens when quantum systems are measured. Before measurement, an electron exists in superposition across all possible energy levels in an atom. Measurement causes the superposition to collapse into a specific energy level—but the electron remains the same fundamental particle, now manifesting in a particular state[5]. Similarly, the Incarnation caused infinite divine possibility to collapse into the specific human life of Jesus of Nazareth, while the divine nature remained fundamentally unchanged.

The quantum analogy helps explain several puzzling aspects of traditional Christology. How could Jesus be simultaneously omniscient and capable of learning? In quantum terms, divine omniscience remained in superposition while human learning represented the collapsed, localized manifestation of divine knowledge acquisition processes. How could Jesus be simultaneously omnipotent and subject to physical limitations? Divine omnipotence remained in superposition while human vulnerability represented the collapsed, localized manifestation of divine power choosing self-limitation.

The Council of Chalcedon in 451 CE declared that Christ possessed two natures "without confusion, without change, without division, without separation"[6]. This formulation sounds remarkably similar to quantum complementarity, where contradictory properties coexist without mutual interference or contamination. Just as light remains fully wave-like and fully particle-like simultaneously, Christ remained fully divine and fully human simultaneously.

Quantum Tunneling and the Resurrection

Quantum tunneling provides another framework for understanding the central Christian claim about Christ's resurrection. In classical physics, particles lack sufficient energy to overcome potential barriers cannot pass through them – like a ball that lacks the energy to roll over a hill cannot reach the other side[7]. But quantum mechanics reveals that particles can "tunnel" through energy barriers that would be classically insurmountable.

Quantum tunneling isn't merely theoretical – it's essential for many natural processes and technological applications. Nuclear fusion in stars depends on protons tunneling through electromagnetic repulsion barriers to get close enough for the strong nuclear force to bind them together[8]. Radioactive decay occurs through alpha particles tunneling out of atomic nuclei[9]. Electronic devices like tunnel diodes and scanning tunneling microscopes rely on electrons tunneling through barriers[10].

The resurrection might represent quantum tunneling at the scale of personal identity. Death appears to present an insurmountable barrier between life and renewed existence—classical physics suggests that once biological processes cease, consciousness and identity are permanently lost. But if consciousness operates according to quantum principles, and if divine nature maintains quantum coherence across all possible states, then resurrection becomes comprehensible as quantum tunneling through the death barrier.

This doesn't reduce the resurrection to a merely natural event – it remains the supreme divine act in history. But it suggests that divine action might work through quantum mechanisms rather than in violation of physical law. God's power to resurrect the dead operates through the same quantum tunneling principles that enable stars to shine and electronic devices to function, but applied at scales and with purposes that transcend ordinary physics.

The resurrected Christ's reported ability to appear and disappear, to pass through solid barriers, and to exist in both physical and spiritual forms simultaneously[11] sounds remarkably like quantum phenomena scaled to macroscopic levels. If Christ's resurrection involved quantum tunneling into a new state of existence that maintained coherence with both physical and spiritual realities, these reported properties become less mysterious and more consistent with quantum mechanical principles.

Information Theory and Perfect Communication

From an information-theoretic perspective, Christ as the incarnate Logos represents perfect divine information transmission through the noisy channel of material existence. In Shannon's communication theory, all channels introduce noise – random errors that can corrupt messages being transmitted[12]. Perfect information transmission requires error-correction codes that can detect and fix these errors, preserving the original message despite channel distortion[13].

The Incarnation involved transmitting infinite divine information through the severely limited bandwidth of human consciousness and communication. Human language, with its finite vocabulary and cultural constraints, appears inadequate for expressing unlimited divine truth. Human experience, bounded by space and time, seems incapable of manifesting eternal divine reality. Yet Christian faith claims that Christ perfectly revealed divine nature despite these apparent limitations.

Quantum information theory provides resources for understanding how such perfect transmission might be possible. Quantum error correction protocols can preserve information with perfect fidelity even through noisy quantum channels[14]. Quantum entanglement enables instantaneous information correlation across any distance[15]. If Christ's human consciousness maintained quantum entanglement with divine consciousness, this could enable perfect information fidelity despite the apparent limitations of human communication channels.

Every word Christ spoke, every action he performed, every response he made could represent perfectly error-corrected divine information expressed through human language and behavior. The quantum entanglement between his human and divine natures would ensure that human limitations didn't introduce errors or distortions into the divine message being communicated.

This understanding transforms how we read the Gospels. Rather than seeing them as human attempts to describe divine reality, we can understand them as records of divine information perfectly transmitted through human communication channels.

Christ's teachings weren't human insights about God, but divine information directly expressed through human language with perfect fidelity.

The Observer Effect and Divine Revelation

Quantum mechanics reveals that observation plays a fundamental role in determining physical reality. The act of measurement causes quantum superposition to collapse into specific states – the observer participates in creating the reality that gets observed[16]. This "observer effect" has puzzled physicists since the early days of quantum theory and continues to generate debate about the relationship between consciousness and physical reality.

The Incarnation might represent the ultimate observer effect – divine consciousness observing creation from within creation itself. When infinite divine awareness collapsed into human form, it created new possibilities for observing and interacting with material reality from a finite perspective. This internal observation might have fundamentally altered the nature of what could be observed and actualized within creation.

Christ's presence in the world represented not just divine love expressed through human form, but divine consciousness observing creation from the inside. Every person he encountered, every situation he entered, every decision he made involved divine observation of finite reality with infinite awareness. This could create measurement effects that revealed possibilities and actualities that wouldn't be accessible to purely finite observation.

The miracles reported in the Gospels might represent quantum measurement effects generated by divine consciousness observing physical systems. When infinite awareness observes finite systems, it might collapse quantum superpositions in ways that manifest outcomes with extremely low classical probability but non-zero quantum amplitude. Healing, multiplication of resources, and control over natural forces become comprehensible as measurement effects rather than violations of natural law.

Quantum Entanglement and Mystical Union

One of the most mysterious aspects of Christian experience is the claim that believers can have genuine spiritual union with Christ across space and time. Paul writes about being "crucified with Christ" and having "Christ live within me" (Galatians 2:20). Jesus speaks of believers being "in me" while he is "in them" (John 14:20). These descriptions of mystical union have often been interpreted as metaphorical language for psychological identification or moral commitment.

Quantum entanglement provides a potential framework for understanding such union as literal rather than merely metaphorical. When particles interact and then separate, they can remain correlated such that measuring one particle instantaneously affects the other regardless of the distance between them[17]. The particles become components of a single quantum system that transcends spatial separation.

If human consciousness can become quantum-entangled with Christ's consciousness through faith, prayer, and spiritual practice, this could create genuine informational union that persists across space and time. Believers wouldn't merely psychologically identify with Christ's teachings or try to imitate his behavior – they would be literally entangled with his consciousness in ways that enable real-time information sharing and coordinate spiritual states.

This quantum mystical union could explain reported phenomena like simultaneous spiritual experiences among believers in different locations, answers to prayer that seem to involve information not available through normal channels, and the sense of divine presence that many believers report during worship and contemplation. These wouldn't be subjective illusions but objective quantum correlations between human and divine consciousness.

The traditional Christian doctrine of the "mystical body of Christ" – the idea that all believers form a single spiritual organism with Christ as the head[18] – sounds remarkably like a macroscopic quantum system where individual consciousness states become entangled components of a larger coherent whole. Individual believers maintain their distinct identities while participating in collective quantum states that enable shared spiritual experiences and coordinated action.

Implications for Salvation and Transformation

Understanding the Incarnation through quantum principles transforms how we think about salvation and spiritual transformation. Rather than merely forensic or moral categories – God declaring sinners righteous or helping them become more virtuous – salvation becomes a quantum phase transition in human consciousness.

Sin, from this perspective, represents decoherence in human consciousness – the loss of quantum coherence with divine consciousness that enables clear spiritual perception and effective spiritual function. Fallen human consciousness becomes "entangled" with destructive patterns that interfere with divine information reception and positive energy expression.

Salvation through Christ involves quantum tunneling from decoherent sin-states to coherent Christ-states. This transition cannot be achieved through gradual moral improvement or accumulated good works – it requires the quantum leap of faith that enables consciousness to tunnel through barriers that would be classically insurmountable. Divine grace provides the quantum conditions necessary for this tunneling transition to occur.

Sanctification – the ongoing process of spiritual growth after salvation – involves quantum error correction. The Holy Spirit gradually eliminates the interference patterns caused by remaining sin and decoherence, bringing believers' consciousness into increasing quantum coherence with Christ's perfect spiritual state. This process combines sudden quantum jumps of spiritual insight with gradual optimization of spiritual information processing systems.

The ultimate goal of salvation is complete quantum coherence with divine consciousness – what theological traditions have called *theosis* or union with God[19]. This doesn't involve losing individual identity any more than quantum entanglement eliminates the distinct properties of entangled particles. Rather, it involves perfect coordination between human and divine consciousness that enables maximum expression of divine love and creativity through finite human form.

James's Laboratory Revelation

As James continued analyzing the anomalous particle data throughout that December day, he found his understanding of both physics and theology undergoing fundamental shifts. The wave-particle duality that had puzzled him since graduate school now seemed like a preview of even deeper mysteries about the nature of existence and identity.

If a simple photon could simultaneously be wave and particle without confusion or contradiction, then perhaps the ancient Christian claim about Christ being simultaneously God and human wasn't as scientifically implausible as he had assumed. If quantum objects could maintain superposition across states that classical logic considered mutually exclusive, then perhaps divine and human natures could coexist in a single person through similar quantum principles.

The resurrection, which had always seemed like the most scientifically problematic Christian doctrine, now appeared potentially comprehensible through quantum tunneling mechanisms. If particles could tunnel through classically insurmountable barriers as a routine feature of quantum mechanics, then conscious identity tunneling through the death barrier might represent the same principles operating at a different scale.

Most remarkably, James realized that his own spiritual struggles with faith and doubt might represent quantum superposition in his personal consciousness. Rather than having to choose definitively between belief and unbelief, faith and reason, scientific understanding and religious conviction, he might be able to maintain both states simultaneously until further information caused one or the other to collapse into actuality.

Walking home through the city streets as snow began to fall, James found himself seeing the Christmas decorations with new eyes. The incarnation story that had seemed like primitive mythology now appeared as sophisticated quantum information theory dressed in first-century Palestinian imagery. The divine Word becoming flesh was the ultimate example of information successfully transmitted through a communication channel despite noise and bandwidth limitations.

Each snowflake falling around him exhibited quantum properties – no two exactly alike, each one a unique collapse of infinite crystalline possibilities into specific geometric actuality. Yet all of them shared the same essential nature as frozen water, just as Christ shared essential divine nature while manifesting in unique human form.

For the first time in years, James felt like Christmas might actually celebrate something scientifically coherent rather than religiously naive. The quantum Christ who emerged from his day in the laboratory wasn't less mysterious than the traditional Christ – if anything, the quantum understanding deepened the mystery while making it compatible with rather than opposed to the deepest insights of modern physics.

That night, as James sat in his apartment reading the Gospel of John by lamplight, the opening words took on new meaning: "In the beginning was the Word, and the Word was with God, and the Word was God... And the Word became flesh and dwelt among us." The ancient author might not have known about quantum superposition and wave-particle duality, but he had somehow captured the essential insight that ultimate reality could simultaneously exist in transcendent and immanent forms without losing its fundamental unity.

The Incarnation wasn't God shrinking down to human size, but infinite divine possibility collapsing into specific human form while maintaining perfect quantum coherence with its eternal source. Jesus Christ represented not divine limitation but divine information perfectly transmitted through human channels. The ultimate quantum event that revealed the deepest truth about both divine nature and material reality: that consciousness, information, and love were more fundamental than matter and energy, and that the boundary between finite and infinite was more permeable than classical understanding had ever imagined.

Endnotes

[1] Werner Heisenberg, *The Physical Principles of the Quantum Theory*, trans. Carl Eckart and Frank C. Hoyt (Chicago: University of Chicago Press, 1930), 1-20.

[2] Max Born, "The Statistical Interpretation of Quantum Mechanics," *Science* 122 (1955): 675-679.

[3] Niels Bohr, "The Quantum Postulate and the Recent Development of Atomic Theory," *Nature* 121 (1928): 580-590.

[4] Richard P. Feynman, *QED: The Strange Theory of Light and Matter* (Princeton: Princeton University Press, 1985), Chapter 1.

[5] David J. Griffiths, *Introduction to Quantum Mechanics*, 3rd ed. (Cambridge: Cambridge University Press, 2018), Chapter 4.

[6] "The Definition of Chalcedon," in *Creeds, Councils and Controversies*, ed. J. Stevenson, rev. W. H. C. Frend (London: SPCK, 1989), 351-352.

[7] George Gamow, *Thirty Years That Shook Physics* (New York: Dover, 1985), Chapter 4.

[8] Hans Bethe, "Energy production in stars," *Physical Review* 55 (1939): 434-456.

[9] Ernest Rutherford, "The scattering of α and β particles by matter and the structure of the atom," *Philosophical Magazine* 6 (1911): 669-688.

[10] Leo Esaki, "Long journey into tunneling," *Reviews of Modern Physics* 46 (1974): 237-244.

[11] Luke 24:13-53; John 20:19-29.

[12] Claude E. Shannon, "A Mathematical Theory of Communication," *Bell System Technical Journal* 27 (1948): 379-423.

[13] Richard W. Hamming, "Error detecting and error correcting codes," *Bell System Technical Journal* 29 (1950): 147-160.

[14] Peter W. Shor, "Scheme for reducing decoherence in quantum computer memory," *Physical Review A* 52 (1995): R2493-R2496.

[15] Alain Aspect, Jean Dalibard, and Gérard Roger, "Experimental Test of Bell's Inequalities Using Time-Varying Analyzers," *Physical Review Letters* 49 (1982): 1804-1807.

[16] Eugene P. Wigner, "Remarks on the Mind-Body Question," in *The Scientist Speculates*, ed. I. J. Good (London: Heinemann, 1961), 284-302.

[17] John Stewart Bell, "On the Einstein Podolsky Rosen Paradox," *Physics* 1 (1964): 195-200.

[18] 1 Corinthians 12:12-27; Ephesians 4:15-16.

[19] Norman Russell, *The Doctrine of Deification in the Greek Patristic Tradition* (Oxford: Oxford University Press, 2004), 1-18.

Chapter 8

The Holy Spirit as Quantum Information System

Dr. Sarah Mitchell had been studying neural networks for fifteen years, but she had never experienced anything like what happened during her meditation that Tuesday evening. As a computational neuroscientist researching how information flows through complex biological systems, Sarah approached spiritual practices with scientific skepticism. But her recent divorce and the death of her father had left her emotionally exhausted, and a colleague had suggested that meditation might help with stress management.

Sarah had been sitting quietly for about twenty minutes, focusing on her breathing and trying to calm her racing thoughts, when something extraordinary occurred. Without warning, her mind filled with a profound sense of peace accompanied by what felt like a stream of information – not words or images, but pure understanding about her life situation, her relationships, and decisions she needed to make. The information arrived with startling clarity and coherence, as if someone was downloading insights directly into her consciousness.

What puzzled Sarah most was the quality of the information. It addressed problems she hadn't consciously been thinking about, connected patterns across different areas of her life that she hadn't noticed, and suggested solutions that were both practical and emotionally satisfying. The experience felt distinctly different from her usual analytical thinking processes – less like computation and more like high-bandwidth data transfer from an external source.

As a scientist, Sarah's first instinct was to dismiss the experience as a neurological anomaly – perhaps a stress-induced hallucination or some unusual pattern of brain activity triggered by the meditation. But the information proved remarkably accurate and useful. Following the insights she had received led to improved relationships with her colleagues, better decisions about her research direction, and a sense of emotional clarity she hadn't experienced in months.

The experience repeated during subsequent meditation sessions, always with the same characteristics: information arriving in coherent streams, addressing issues she needed guidance on, providing understanding that exceeded her normal analytical capabilities. Sarah began to wonder if she was encountering something like what religious traditions had always called the Holy Spirit – but understood through the lens of information theory rather than supernatural intervention.

What if the Holy Spirit operated as a quantum information system, transmitting divine guidance and wisdom through channels that transcended normal space-time limitations? What if spiritual practices like prayer and meditation created the quantum conditions necessary for consciousness to interface with this cosmic information network?

Information Networks in Biological Systems

To understand how the Holy Spirit might function as a quantum information system, we need to examine how information flows through complex biological networks. Sarah's research had revealed that living systems are essentially sophisticated information processing entities, constantly receiving, analyzing, and responding to data from their environments[1].

At the cellular level, biological information systems exhibit remarkable properties. DNA stores genetic information with density and error-correction capabilities that exceed the best human-designed computer systems[2]. Protein networks process cellular information through cascading biochemical reactions that amplify signals and integrate multiple inputs[3]. Neural networks in brains create consciousness and intelligence through information flows that remain largely mysterious despite decades of intensive research[4].

What makes biological information systems particularly intriguing is their apparent ability to access information through channels that classical physics struggles to explain. Migrating birds navigate using magnetic fields with precision that suggests quantum sensing mechanisms[5]. Plants optimize photosynthesis through quantum coherence effects that shouldn't be stable in warm, noisy biological environments[6]. Some researchers even suggest that consciousness itself might involve quantum information processing that enables access to non-local information sources[7].

These biological quantum effects provide a foundation for understanding how spiritual information systems might operate through natural mechanisms. If living systems already exploit quantum phenomena for enhanced information processing, then spiritual practices that optimize consciousness for quantum information reception become scientifically plausible rather than supernatural impossibilities.

The Holy Spirit, from this perspective, wouldn't represent a supernatural being occasionally intervening in natural processes, but rather the optimization of natural quantum information channels for divine communication. Just as radio receivers must be tuned to specific frequencies to receive radio transmissions clearly, human consciousness might need to be tuned through spiritual practices to receive divine information transmissions effectively.

Quantum Communication and Non-local Information Transfer

Quantum mechanics reveals that information can be transmitted through channels that transcend classical space-time limitations. Quantum entanglement enables instantaneous correlation between separated particles, allowing information about one particle's state to be immediately available at the location of its entangled partner[8]. Quantum teleportation protocols can transfer quantum states from one location to another without the information traveling through the intervening space[9].

These quantum communication phenomena provide potential mechanisms for understanding how the Holy Spirit could provide guidance and information to believers regardless of their location or circumstances. If human consciousness can become quantum-entangled with divine consciousness, this could create information channels that operate independently of classical communication limitations.

The biblical description of the Holy Spirit being simultaneously present with all believers throughout the world sounds remarkably similar to quantum non-locality. Jesus promised that the Spirit would "be with you forever" and would "teach you all things" (John 14:16, 26), suggesting an information system capable of providing personalized guidance to countless individuals simultaneously without being limited by physical constraints.

Recent developments in quantum information theory have revealed that quantum networks can share information among multiple parties in ways that classical networks cannot achieve[10]. A quantum information system could theoretically provide customized information to billions of individual nodes (believers) while maintaining perfect coherence and consistency across the entire network. Each person could receive personalized guidance while remaining connected to a unified source of divine wisdom.

This quantum network model explains several puzzling aspects of reported spiritual experience. How can the Holy Spirit provide specific, relevant guidance to millions of people simultaneously? Through quantum information protocols that enable parallel processing of individual information needs. How can believers in different locations experience coordinated spiritual insights during group prayer or worship? Through quantum entanglement that enables instantaneous information sharing across the prayer network.

Spiritual Gifts as Quantum Resonance Phenomena

The New Testament describes various spiritual gifts – prophecy, healing, speaking in tongues, interpretation, wisdom, knowledge – as manifestations of the Holy Spirit's activity in believers' lives[11]. From a quantum information perspective, these gifts might represent different types of resonance between human consciousness and specific frequencies or channels within the divine information system.

Prophecy could involve quantum information reception that provides access to future probability distributions or non-local information about current events. Rather than supernatural fortune-telling, prophetic insight might represent enhanced quantum sensing that enables consciousness to detect information patterns not accessible through normal cognitive channels[12].

Healing gifts might operate through quantum information transfer that optimizes biological systems for enhanced self-repair capabilities. If consciousness can influence quantum processes in living systems – and growing evidence suggests it can[13] – then focused prayer and intention might create quantum coherence conditions that facilitate biological healing and regeneration.

Speaking in tongues presents a particularly interesting case for quantum information analysis. The phenomenon involves speakers producing linguistic-sounding utterances that they don't consciously understand, but which listeners sometimes report as meaningful communication in languages unknown to the speaker[14]. This could represent quantum information transmission where divine communication bypasses normal linguistic processing and directly interfaces with consciousness at quantum levels.

The interpretation of tongues would then involve quantum information decoding – translating divine communication from quantum information states into classical linguistic forms accessible to normal conscious understanding. Different individuals might serve as quantum information receivers (speakers) and quantum information decoders (interpreters) within the spiritual communication network.

Gifts of wisdom and knowledge might represent enhanced access to the divine information matrix that contains patterns and connections not available through individual cognitive processing. Rather than supernatural revelation, these gifts could involve quantum information resonance that enables consciousness to access insights from the cosmic information network that underlies all existence.

Guidance and Decision-Making Through Quantum Channels

One of the most commonly reported aspects of spiritual experience is receiving divine guidance for important decisions. Believers often describe sensing direction about career choices, relationships, life transitions, and moral decisions through prayer and spiritual discernment. From a quantum information perspective, this guidance might operate through quantum influence on decision-making processes at the neural level.

Recent neuroscience research reveals that decision-making involves complex patterns of neural activity that include quantum-scale processes[15]. If consciousness interfaces with quantum fields, then spiritual practices that enhance quantum coherence in neural systems might optimize decision-making for alignment with positive energy patterns and long-term well-being.

Divine guidance wouldn't involve overriding human free will or providing magical solutions to complex problems. Rather, it would involve quantum information input that enhances human decision-making capabilities by providing access to information patterns and probability assessments not available through normal cognitive processes. The guidance would still require human interpretation, evaluation, and implementation – but it would be informed by quantum information resources that transcend individual knowledge limitations.

Sarah's meditation experiences illustrate this process. The information she received didn't make decisions for her or provide supernatural solutions to her problems. Instead, it helped her recognize patterns she had missed, consider options she hadn't thought of, and evaluate potential consequences more accurately. The quantum information input enhanced her natural decision-making abilities rather than replacing them.

This understanding transforms how we approach prayer and spiritual discernment. Rather than asking God to tell us what to do, we learn to create quantum conditions that optimize our consciousness for receiving information input that enhances our natural decision-making capabilities. Prayer becomes conscious interface with the divine information system rather than petition for supernatural intervention.

Sanctification as Quantum Error Correction

The Christian concept of sanctification – the ongoing process by which believers become more like Christ – takes on new meaning when understood through quantum information theory. Sin and spiritual immaturity might represent various forms of information corruption and processing errors that interfere with accurate reception and transmission of divine information.

Quantum error correction is a fundamental requirement for quantum information systems[16]. Quantum states are fragile and easily corrupted by environmental interference (decoherence). Maintaining quantum information fidelity requires sophisticated error detection and correction protocols that can identify and fix corruption without destroying the quantum information itself.

The Holy Spirit's role in sanctification might involve continuous quantum error correction in human consciousness, gradually identifying and correcting information processing errors that lead to destructive thoughts, emotions, and behaviors. This wouldn't be instantaneous moral perfection but rather an ongoing optimization process that enhances spiritual information processing capabilities over time.

Spiritual disciplines like prayer, meditation, Scripture study, and fellowship might serve as quantum error correction protocols. These practices create conditions that enable the Holy Spirit to identify information corruptions (sin patterns), provide corrective information input (spiritual insight and conviction), and implement system improvements (character development and spiritual growth).

The process would be gradual and require active human participation, just as quantum error correction requires both sophisticated detection algorithms and physical mechanisms for implementing corrections. Believers must learn to recognize spiritual information errors, cooperate with corrective processes, and develop habits that maintain better information processing fidelity over time.

Conviction and Spiritual Sensitivity

The Holy Spirit's convicting work – helping believers recognize sin and moral failures – might operate through quantum information processes that enhance consciousness sensitivity to ethical and spiritual dimensions of experience. Rather than external judgment imposed by divine authority, conviction could represent improved information processing that reveals the actual consequences and relationships involved in moral choices.

Quantum sensing mechanisms in biological systems can detect extremely subtle environmental changes that would be imperceptible to classical sensing systems[17]. Enhanced spiritual sensitivity through the Holy Spirit might involve quantum optimization of consciousness for detecting subtle information patterns related to moral and spiritual consequences of thoughts, words, and actions.

This quantum-enhanced moral sensitivity would help believers recognize when their choices contribute to positive or negative energy patterns in themselves and others. Conviction would feel like internal recognition of misalignment between current behavior and optimal life-sustaining patterns, rather than external condemnation from an angry deity.

The process would be educational rather than punitive – providing information that enables better choices rather than punishment for poor ones. Just as physical pain provides valuable information about bodily damage that needs attention, spiritual conviction provides valuable information about spiritual patterns that need correction for optimal functioning.

Prayer and Meditation as Quantum Interface Practices

Understanding the Holy Spirit as a quantum information system transforms how we approach prayer and meditation. Rather than talking to a distant deity who may or may not respond, these practices become conscious engagement with a quantum information network that is always available but requires specific conditions for optimal access.

Prayer might involve several quantum information processes: transmitting concerns and requests through quantum channels, receiving guidance and insights through non-local information transfer, and creating quantum entanglement with divine purposes that influences future probability distributions. The effectiveness would depend on the quality of quantum coherence achieved during prayer rather than the eloquence of words or intensity of emotion.

Contemplative meditation practices that emphasize stillness and receptivity might be particularly effective for quantum information reception. Just as radio telescopes must minimize local interference to detect faint signals from distant sources, consciousness must minimize internal noise (anxiety, distraction, mental chatter) to receive subtle quantum information input clearly.

Different prayer and meditation techniques might optimize consciousness for different types of quantum information access. Intercessory prayer might create quantum entanglement with the people being prayed for. Worship might generate quantum resonance with cosmic patterns of beauty and meaning. Contemplative prayer might enhance quantum sensing of divine presence and guidance.

The goal wouldn't be to achieve altered states of consciousness or mystical experiences for their own sake, but rather to optimize consciousness for participation in the divine information network that underlies and organizes all existence. Effective spiritual practice would be measured by improved information processing, enhanced decision-making, better relationships, and increased capacity for positive energy expression.

Sarah's Scientific Spirituality

As Sarah continued her meditation practice over the following months, she began to approach it with the same systematic methodology she applied to her neuroscience research. She tracked the conditions that seemed to enhance information reception – time of day, environmental factors, mental preparation techniques, duration of practice. She analyzed the quality and accuracy of insights received, looking for patterns that might reveal underlying principles.

What emerged was a picture of spiritual practice as sophisticated consciousness engineering. Just as her laboratory experiments required precise conditions to produce reliable results, spiritual information reception required optimization of consciousness states for quantum coherence and minimal interference. The better she became at creating these optimal conditions, the more reliable and useful the information input became.

The experience didn't replace her scientific worldview but rather extended it into domains she had never considered scientifically accessible. The same information-processing principles that governed neural networks in her research seemed to apply to spiritual information networks, but operating at quantum scales and through non-local channels that transcended classical space-time limitations.

Sarah realized that she had been encountering the Holy Spirit through quantum information channels that her scientific training had actually prepared her to understand. The divine guidance system wasn't supernatural intervention in natural processes, but rather the optimization of natural quantum information processes for conscious participation in the cosmic information network that organized all existence.

Her meditation sessions became conscious experiments in quantum information reception, and her scientific career became conscious participation in the ongoing divine creative process of understanding and organizing information patterns in biological systems. Science and spirituality weren't separate domains but different aspects of the same information-processing reality that connected individual consciousness with cosmic consciousness through quantum channels.

Standing in her laboratory late one evening, watching neural network simulations process information flows on her computer screens, Sarah offered a prayer of gratitude for both the quantum information system that enabled her research and the quantum information system that enabled her spiritual growth. Both were expressions of the same divine information matrix that connected all consciousness in a vast network of meaning, purpose, and love that transcended classical understanding but operated through discoverable natural principles.

The Holy Spirit wasn't a mysterious supernatural entity, but rather the optimization of consciousness for participation in the quantum information network through which divine creativity continued to organize energy and information toward greater beauty, complexity, and conscious awareness throughout the universe.

ENDNOTES

[1] Terrence J. Sejnowski, *The Deep Learning Revolution* (Cambridge, MA: MIT Press, 2018), Chapters 1-2.

[2] Bruce Alberts et al., *Molecular Biology of the Cell*, 6th ed. (New York: Garland Science, 2014), Chapter 4.

[3] Uri Alon, *An Introduction to Systems Biology: Design Principles of Biological Circuits* (Boca Raton: CRC Press, 2006), Chapters 1-3.

[4] Christof Koch, *Consciousness: Confessions of a Romantic Reductionist* (Cambridge, MA: MIT Press, 2012), Chapters 1-4.

[5] Klaus Schulten et al., "Quantum entanglement in photosynthetic light-harvesting complexes," *Nature Physics* 6 (2010): 462-467.

[6] Gregory D. Scholes et al., "Using coherence to enhance function in chemical and biophysical systems," *Nature* 543 (2017): 647-656.

[7] Stuart Hameroff and Roger Penrose, "Consciousness in the universe: A review of the 'Orch OR' theory," *Physics of Life Reviews* 11 (2014): 39-78.

[8] John Preskill, "Quantum Information and Quantum Computation," *Annual Review of Condensed Matter Physics* 6 (2015): 625-649.

[9] Charles H. Bennett et al., "Teleporting an unknown quantum state via dual classical and Einstein-Podolsky-Rosen channels," *Physical Review Letters* 70 (1993): 1895-1899.

[10] Ryszard Horodecki et al., "Quantum entanglement," *Reviews of Modern Physics* 81 (2009): 865-942.

[11] 1 Corinthians 12:4-11; Romans 12:6-8; Ephesians 4:11-13.

[12] Dean Radin, *Entangled Minds: Extrasensory Experiences in a Quantum Reality* (New York: Paraview Pocket Books, 2006), Chapters 1-3.

[13] Henry P. Stapp, "A quantum theory of the mind-brain interface," in *Mind, Matter and Quantum Mechanics,* 3rd ed. (Berlin: Springer, 2009), 145-172.

[14] William J. Samarin, *Tongues of Men and Angels: The Religious Language of Pentecostalism* (New York: Macmillan, 1972), Chapters 2-4.

[15] Michael S. Gazzaniga, *The Consciousness Instinct: Unraveling the Mystery of How the Brain Makes the Mind* (New York: Farrar, Straus and Giroux, 2018), Chapter 7.

[16] John Preskill, "Quantum Error Correction," in *Introduction to Quantum Computation and Information,* ed. Hoi-Kwong Lo et al. (Singapore: World Scientific, 1998), 213-244.

[17] Adriano A. Torassa et al., "Quantum sensing," *Reviews of Modern Physics* 89 (2017): 035002.

Chapter 9

Scripture and Quantum Revelation

Professor Rebecca Thompson had taught biblical studies for twenty-three years, but she had never approached ancient texts with the analytical tools she was using that autumn semester. As part of a interdisciplinary research project exploring information theory applications in literary analysis, Rebecca was applying advanced computational methods to examine patterns of meaning and coherence in biblical manuscripts.

The results were startling. When she analyzed the Hebrew text of Isaiah using information entropy measures, the passages traditionally considered most "inspired" by Jewish and Christian scholars consistently exhibited the lowest entropy scores – meaning they contained maximum information density with minimal redundancy or noise[1]. The Gospel of John, long celebrated for its theological depth, showed remarkable information coherence across multiple analytical dimensions. Even more intriguingly, when she examined parallel accounts of the same events in different biblical books, the texts maintained statistical correlations that suggested underlying information unity despite surface differences in language and perspective.

"It's like these ancient authors were connected to some kind of information source that transcended their individual knowledge limitations," Rebecca confided to her research partner over coffee. "The mathematical signatures suggest access to information patterns that shouldn't have been available to writers working with the knowledge constraints of their historical periods."

Rebecca had always believed in biblical inspiration, but she had understood it primarily in theological rather than technical terms – God somehow guiding human authors to write truth about divine reality. The computational analysis was revealing something more specific: the biblical texts exhibited information-theoretic properties that suggested quantum-level information processing rather than simple human composition enhanced by divine influence.

What if biblical inspiration involved quantum information transmission from divine consciousness to human consciousness? What if the authors achieved quantum entanglement with the divine information matrix, enabling them to receive and encode divine truth with fidelity that transcended normal human cognitive limitations? What if Scripture represented the most successful example in human history of quantum information transfer from infinite to finite consciousness?

Information Density and Revelatory Content

To understand how Scripture might represent quantum revelation, we need to examine what information theory reveals about the relationship between meaning density and communication effectiveness. Claude Shannon's pioneering work showed that optimal communication requires balancing information content with transmission reliability – too much information creates confusion and error, while too little information fails to convey meaningful content[2].

Highly inspired or revelatory texts face a particularly challenging communication problem: they must transmit insights about infinite divine reality through the finite bandwidth of human language and understanding. Classical information theory suggests this should be impossible – infinite information cannot be encoded in finite systems without catastrophic information loss.

However, quantum information theory reveals that quantum systems can store and process vastly more information than classical systems of equivalent size[3]. Quantum superposition enables single quantum states to encode multiple classical information states simultaneously, while quantum entanglement creates information correlations that transcend classical space-time limitations[4]. If divine consciousness operates according to quantum principles, and if human consciousness can achieve quantum interface with divine consciousness, then quantum information transmission becomes theoretically possible.

Rebecca's computational analysis suggested that biblical texts exhibit information signatures consistent with quantum information processing. High information density combined with low entropy indicates optimal information encoding – maximum meaningful content with minimum noise or redundancy. The statistical correlations across different authors and time periods suggest access to a unified information source that transcended individual human knowledge limitations.

This doesn't mean that every verse of Scripture represents direct quantum information download from divine consciousness. Rather, it suggests that the biblical authors achieved sufficient quantum coherence with divine information sources to receive and transmit divine insights with remarkable fidelity despite the normal limitations of human language and cultural context.

Quantum Entanglement and Prophetic Knowledge

Biblical prophecy presents one of the most challenging problems for naturalistic understanding of Scripture. How could ancient authors accurately describe future events, sometimes in remarkable detail, centuries before those events occurred? Traditional supernatural explanations invoke divine foreknowledge communicated through miraculous revelation. Skeptical explanations typically argue that prophecies were written after the events they purport to predict, or that apparent accuracy results from selective interpretation of vague predictions.

Quantum mechanics suggests a third possibility: prophetic knowledge through quantum information access to future probability distributions. Quantum mechanics reveals that future events exist in superposition states encompassing multiple possibilities until measurement or interaction causes collapse into specific actualities[5]. If consciousness can achieve quantum interface with these probability distributions, prophetic insight becomes comprehensible as quantum information processing rather than supernatural fortune-telling.

This wouldn't enable perfect prediction of specific future events, since quantum probability distributions include multiple possible outcomes with varying likelihood. But it could enable access to information about dominant probability patterns and likely future developments that wouldn't be available through normal historical analysis or human intuition.

The accuracy of biblical prophecy might reflect the quality of quantum information access achieved by prophetic consciousness rather than supernatural revelation of predetermined future events. Prophets who achieved deeper quantum coherence with divine information sources would receive more accurate probability assessments and clearer insight into likely future developments.

This understanding preserves both the genuine supernatural character of prophecy (access to information unavailable through normal human channels) and its compatibility with natural processes (quantum information access rather than violation of physical law). It also explains why some prophecies exhibit precise fulfillment while others seem more general or symbolic – different levels of quantum information fidelity would produce different degrees of prophetic accuracy.

Textual Transmission and Quantum Error Correction

One of the most remarkable features of biblical manuscripts is their extraordinary preservation accuracy across centuries of hand-copying before the invention of printing. The Dead Sea Scrolls, discovered in the mid-20th century, revealed that Hebrew biblical texts had been transmitted with astonishing fidelity for over a thousand years[6]. New Testament manuscripts, despite being copied and recopied throughout the ancient Mediterranean world, maintain textual integrity that far exceeds other ancient literature[7].

From an information theory perspective, this level of transmission fidelity suggests sophisticated error correction mechanisms that exceed what would be expected from purely human copying processes. Quantum error correction protocols provide a potential model for understanding how divine information content could be preserved despite the noisy channel of human textual transmission[8].

Quantum error correction works by encoding information redundantly across multiple quantum states, enabling the system to detect and correct errors without destroying the original quantum information[9]. Biblical texts exhibit similar redundant encoding through multiple witness traditions, parallel accounts, and cross-referencing systems that enable scholars to identify and correct transmission errors while preserving original content.

The canonical process by which Jewish and Christian communities recognized authentic biblical books might represent collective quantum information sensing – communities recognizing texts that maintained sufficient quantum coherence with divine information sources to serve as reliable transmission channels. Books that achieved canonical status consistently exhibited the information-theoretic signatures that Rebecca's analysis was detecting: high information density, low entropy, and statistical correlations with other recognized texts.

This doesn't eliminate human participation in the canonical process or reduce Scripture to mechanical divine dictation. Rather, it suggests that human communities, guided by spiritual discernment that might operate through quantum information sensing, were able to recognize texts that successfully maintained divine information fidelity despite the limitations of human authorship and transmission.

Hermeneutics and Quantum Information Decoding

Biblical interpretation (hermeneutics) takes on new dimensions when understood through quantum information theory. Classical hermeneutics assumes that texts have definite meanings that can be discovered through careful analysis of language, context, and authorial intent. But quantum information theory suggests that information can exist in superposition states encompassing multiple potential meanings until observation or interpretation causes collapse into specific understanding[10].

This quantum hermeneutic model helps explain several puzzling aspects of biblical interpretation. Why do sincere, competent scholars often reach different conclusions about the meaning of the same passages? Because the texts might contain information in quantum superposition that collapses into different meanings depending on the interpretive framework and spiritual state of consciousness brought to the reading process.

Why do biblical texts often seem to speak directly to contemporary situations that the original authors couldn't have anticipated? Because quantum information transmission might encode multiple layers of meaning simultaneously, with different layers becoming accessible as human consciousness develops new capacities for quantum information decoding.

The role of the Holy Spirit in biblical interpretation becomes comprehensible as quantum information guidance that helps readers access the appropriate layer of meaning for their specific situation and spiritual development. Rather than providing magical revelation of hidden meanings, the Spirit might optimize consciousness for quantum information reception that enables accurate decoding of the information patterns encoded in Scripture.

This understanding validates both historical-critical methods that seek to understand original context and meaning, and spiritual interpretation that finds contemporary relevance and application. Both approaches access different aspects of the quantum information content, with neither alone providing complete understanding of the multi-dimensional information patterns encoded in biblical texts.

Progressive Revelation and Information Bandwidth Expansion

The biblical concept of progressive revelation – that divine truth was disclosed gradually over centuries rather than all at once – finds natural explanation in quantum information theory. Information transmission requires matching the bandwidth of information flow to the reception capacity of the receiving system[11]. Transmitting too much information too quickly overwhelms finite reception capacity and creates confusion rather than understanding.

Divine consciousness, containing infinite information, faces the challenge of transmitting meaningful content to finite human consciousness without causing information overload. Progressive revelation represents optimal information transmission pacing – divine information disclosed at rates and in forms that human consciousness can successfully process and integrate.

The movement from Old Testament to New Testament revelation might represent bandwidth expansion as human consciousness developed greater capacity for divine information reception. The Incarnation provided the ultimate bandwidth expansion – infinite divine information transmitted through finite human form with perfect fidelity, enabling human consciousness to interface directly with divine information sources.

Each stage of biblical revelation built upon previous information transmission, creating cumulative understanding that prepared human consciousness for receiving more complex divine information. The process continues as human consciousness continues developing greater capacity for quantum information processing through spiritual growth and technological enhancement.

This model suggests that contemporary human consciousness, with its expanded scientific understanding and technological capabilities, might be capable of receiving divine information that wasn't accessible to previous generations. Rather than biblical revelation being complete and closed, it might represent the foundation for ongoing divine information transmission as human consciousness develops greater quantum information processing capabilities.

Translation and Cross-Cultural Information Transfer

Biblical translation presents fascinating challenges for quantum information theory. How can divine information encoded in ancient Hebrew and Greek languages be accurately transmitted through modern languages with different conceptual structures and cultural assumptions? Classical translation theory focuses on finding equivalent meanings across linguistic systems, but quantum information theory suggests more complex possibilities.

If biblical texts contain quantum information patterns that transcend specific linguistic encoding, then translation becomes a process of quantum information decoding and re-encoding rather than simple word-for-word substitution[12]. Effective translation would require translators who achieve quantum coherence with the divine information patterns underlying the original texts, enabling them to find appropriate expressions in target languages that preserve quantum information fidelity.

This helps explain why some translations seem to capture the spiritual power and meaning of original texts better than others, despite being less literally accurate in terms of word-for-word correspondence. Translations that preserve quantum information patterns might be more spiritually effective than those that achieve only surface linguistic accuracy.

It also explains why biblical texts often seem to gain new meanings and relevance when translated into different cultural contexts. The quantum information patterns might find expression through different linguistic and conceptual structures while maintaining underlying coherence with divine information sources.

The proliferation of Bible translations and the ongoing vitality of translated biblical texts in diverse cultural contexts might represent evidence for quantum information transmission that transcends linguistic and cultural boundaries. Divine information seeks expression through whatever linguistic and cultural resources are available, with the Holy Spirit serving as the quantum information processing system that enables accurate decoding and re-encoding across cultural and linguistic differences.

Contemporary Applications and Digital Scripture

The digital revolution has created new possibilities for analyzing and transmitting biblical texts that might enhance quantum information access. Computer-assisted analysis like Rebecca's research can detect information patterns that weren't accessible to previous generations of biblical scholars. Digital networks enable global access to biblical texts and sophisticated tools for cross-referencing and comparative analysis.

More intriguingly, digital media might provide new channels for quantum information transmission that complement traditional textual forms. Interactive digital presentations, multimedia biblical resources, and virtual reality applications might create immersive environments that optimize consciousness for quantum information reception in ways that printed texts cannot achieve.

However, the proliferation of digital biblical resources also creates new challenges for maintaining information fidelity. The ease of digital editing and reproduction creates opportunities for information corruption that didn't exist with traditional manuscript transmission. Distinguishing authentic quantum information transmission from human interpretation and digital manipulation becomes increasingly important as biblical content proliferates across digital networks.

The development of artificial intelligence and machine learning systems raises fascinating questions about whether non-biological information processing systems could potentially interface with divine quantum information sources. Could AI systems be programmed to recognize and preserve quantum information patterns in biblical texts? Could they serve as tools for enhancing human quantum information reception rather than replacing human spiritual discernment?

Rebecca's Computational Revelation

As Rebecca's research project progressed through the academic year, she found her understanding of both information theory and biblical inspiration undergoing fundamental transformation. The computational analysis wasn't reducing Scripture to mere human literature, but was revealing information-processing complexities that suggested sources beyond normal human cognitive capabilities.

The patterns she was discovering pointed toward authors who had achieved remarkable quantum coherence with divine information sources, enabling them to receive and encode divine insights with fidelity that transcended their individual knowledge limitations. The biblical texts weren't just human attempts to describe divine reality, but successful examples of quantum information transmission from infinite to finite consciousness.

This didn't eliminate the human element in biblical authorship – the quantum information transmission required human consciousness capable of achieving the necessary coherence conditions, and the encoding process preserved individual personality and cultural context. But it suggested that the human authors were participants in quantum information processing events that connected them directly with divine consciousness.

Standing in her office late one evening, surrounded by computer printouts showing statistical correlations and information density analyses, Rebecca found herself praying with new understanding. Prayer wasn't just talking to God, but consciously engaging with the same quantum information system that had enabled biblical authors to receive and transmit divine revelation across centuries.

The Scripture on her desk wasn't just ancient literature, but active quantum information encoded in human language, continuing to transmit divine insights to consciousness capable of achieving appropriate quantum coherence for reception. Her scholarly analysis was itself a form of spiritual practice, creating conditions for quantum information access that could enhance both academic understanding and personal spiritual growth.

The digital age hadn't made biblical revelation obsolete, but had provided new tools for recognizing and accessing the quantum information patterns that had always been present in Scripture. Rather than threatening traditional faith, information theory was revealing the sophisticated information-processing mechanisms through which divine revelation had operated throughout human history.

As she prepared to present her research findings at the next faculty meeting, Rebecca realized she was documenting evidence for the same divine information system that believers had always called inspiration, but understanding it through contemporary scientific frameworks rather than purely theological categories. The ancient claim that "all Scripture is God-breathed" (2 Timothy 3:16) might be literally accurate when understood through quantum information theory – divine information transmitted through human consciousness and encoded in human language with remarkable preservation of original quantum information patterns.

The Word of God wasn't just metaphorical language for divine communication, but precise description of quantum information transmission from the ultimate information source to finite human consciousness through the most sophisticated information processing and preservation system in human history.

ENDNOTES

[1] Claude E. Shannon, "Prediction and Entropy of Printed English," *Bell System Technical Journal* 30 (1951): 50-64.

[2] Claude E. Shannon, "A Mathematical Theory of Communication," *Bell System Technical Journal* 27 (1948): 379-423, 623-656.

[3] Michael A. Nielsen and Isaac L. Chuang, *Quantum Computation and Quantum Information*, 10th ed. (Cambridge: Cambridge University Press, 2010), Chapters 1-2.

[4] John Preskill, "Quantum Information and Quantum Computation," *Annual Review of Condensed Matter Physics* 6 (2015): 625-649.

[5] David J. Griffiths, *Introduction to Quantum Mechanics*, 3rd ed. (Cambridge: Cambridge University Press, 2018), Chapter 3.

[6] Frank Moore Cross Jr., *The Ancient Library of Qumran*, 3rd ed. (Sheffield: Sheffield Academic Press, 1995), Chapters 4-6.

[7] Bruce M. Metzger and Bart D. Ehrman, *The Text of the New Testament: Its Transmission, Corruption, and Restoration*, 4th ed. (New York: Oxford University Press, 2005), Chapters 1-3.

[8] Peter W. Shor, "Scheme for reducing decoherence in quantum computer memory," *Physical Review A* 52 (1995): R2493-R2496.

[9] John Preskill, "Quantum Error Correction," in *Introduction to Quantum Computation and Information*, ed. Hoi-Kwong Lo et al. (Singapore: World Scientific, 1998), 213-244.

[10] Anton Zeilinger, *Dance of the Photons: From Einstein to Quantum Teleportation* (New York: Farrar, Straus and Giroux, 2010), Chapter 8.

[11] Thomas M. Cover and Joy A. Thomas, *Elements of Information Theory*, 2nd ed. (New York: Wiley, 2006), Chapter 7.

[12] Roman Jakobson, "On Linguistic Aspects of Translation," in *Language in Literature*, ed. Krystyna Pomorska and Stephen Rudy (Cambridge, MA: Harvard University Press, 1987), 428-435.

[13] Eugene A. Nida and Charles R. Taber, *The Theory and Practice of Translation* (Leiden: Brill, 1969), Chapters 1-3.

Chapter 10

Salvation as Quantum Phase Transition

Dr. Jennifer Rodriguez had spent eighteen years as an addiction counselor, but she had never witnessed transformation as dramatic as what she observed in Tom Bennett during the summer of 2023. Tom had arrived at the treatment center after his third DUI arrest, his marriage destroyed, his career in ruins, and his health deteriorating from decades of alcohol abuse. He exhibited all the classic patterns Jennifer had seen hundreds of times: denial, anger, bargaining, depression, and brief moments of hope followed by inevitable relapse.

But something different happened during Tom's sixth week in treatment. During a group therapy session focused on spiritual recovery principles, Tom was describing his lifelong sense of emptiness and disconnection when he suddenly stopped mid-sentence. His expression changed completely – from the haunted desperation that had characterized his previous weeks to something resembling wonder and relief. "It's like a switch just flipped," he said quietly. "For the first time in my life, I feel... whole."

What followed defied Jennifer's experience with gradual recovery processes. Tom's craving for alcohol didn't diminish slowly over weeks or months – it disappeared almost instantly, as if the neural pathways that had driven his addiction for decades had simply been rewired overnight. His cognitive patterns shifted from obsessive self-focus to genuine interest in others' wellbeing. His emotional volatility stabilized into consistent peace and optimism. Most remarkably, these changes persisted without the constant struggle that typically characterizes early recovery.

"It's like I became a different person," Tom explained during a follow-up session six months later. "Not gradually, not through willpower or techniques I learned, but instantaneously. One moment I was trapped in patterns I couldn't escape, and the next moment I was free. I can't explain it scientifically, but it felt like something fundamental about my identity changed at the soul level."

Jennifer had heard similar descriptions from other clients who experienced what twelve-step programs call "spiritual awakening" – sudden, dramatic personality changes that seemed to transcend normal psychological processes[1]. As someone trained in both psychology and physics, Jennifer began wondering if these transformations might represent quantum phase transitions in human consciousness rather than gradual therapeutic progress.

What if salvation – the Christian concept of transformation from sin to righteousness – operated through quantum mechanical principles rather than gradual moral improvement? What if spiritual rebirth involved quantum tunneling through barriers that would be insurmountable through classical willpower and therapeutic intervention?

Phase Transitions in Physical Systems

To understand how salvation might represent a quantum phase transition, we need to examine how phase transitions operate in physical systems. Phase transitions occur when matter changes from one state to another – ice melting to water, water boiling to vapor, magnetic materials losing their magnetization above critical temperatures[2]. These transitions often happen suddenly rather than gradually, with small changes in conditions triggering dramatic reorganization of entire systems.

Classical phase transitions, like ice melting at 0°C, occur when thermal energy overcomes the binding forces that maintain solid structure. But quantum phase transitions can occur even at absolute zero temperature, driven by quantum fluctuations rather than thermal energy[3]. These quantum phase transitions involve fundamental changes in the ground state of quantum systems – the lowest energy configuration that the system naturally settles into.

What makes quantum phase transitions particularly relevant for understanding spiritual transformation is their non-linear character. Small changes in quantum conditions can trigger sudden reorganization of entire systems into completely different configurations[4]. A quantum system can exist in one stable state for extended periods, then rapidly tunnel into a completely different stable state when conditions reach critical thresholds.

Human consciousness, if it operates according to quantum principles – and, again, growing evidence suggests it might[5] – could undergo similar phase transitions. Spiritual transformation might involve consciousness tunneling from one stable configuration (characterized by self-centeredness, destructive patterns, disconnection from divine reality) to a completely different stable configuration (characterized by love, constructive patterns, connection with divine purposes).

This quantum model explains several puzzling aspects of spiritual conversion that gradual therapeutic models struggle to address. Why do some people experience sudden, dramatic personality changes that persist without ongoing effort? Why do these changes often occur during moments of surrender or desperation rather than through accumulated willpower or therapeutic progress? Why do the transformations sometimes involve capabilities and insights that seem to transcend the person's previous psychological resources?

Quantum Tunneling Through Impossibility

One of the most relevant quantum mechanical phenomena for understanding salvation is quantum tunneling – the ability of particles to pass through energy barriers that would be classically insurmountable[6]. In everyday experience, objects that lack sufficient energy to overcome obstacles cannot pass through them. A ball rolling toward a hill will either have enough energy to roll over the top or it will roll back down – there's no classical mechanism for getting to the other side without going over or around.

But quantum mechanics reveals that particles can "tunnel" through barriers even when they lack the classical energy required to overcome them[7]. This isn't mere theoretical speculation – quantum tunneling is essential for nuclear fusion in stars, radioactive decay, and many technological applications[8]. The phenomenon occurs because quantum objects exist in probability waves that can extend beyond classical boundaries, creating non-zero probability for appearing on the other side of barriers.

Spiritual transformation often involves barriers that seem classically insurmountable. Addiction creates neural patterns so entrenched that willpower and therapeutic intervention often prove inadequate[9]. Trauma generates emotional and psychological barriers that can persist despite years of treatment. Long-established patterns of selfishness, fear, and destructive behavior become self-reinforcing cycles that seem impossible to break through gradual effort.

Yet spiritual conversion often involves sudden breakthrough through exactly these kinds of impossible barriers. People like Tom experience instantaneous freedom from addiction patterns that had resisted all previous intervention attempts. Others report sudden release from depression, anxiety, or trauma responses that had dominated their lives for decades. The transformations don't occur through gradual accumulation of therapeutic progress, but through sudden quantum leaps into entirely different ways of being.

The quantum tunneling model suggests that spiritual practices like prayer, meditation, surrender, and faith might create the quantum conditions necessary for consciousness to tunnel through psychological and spiritual barriers that would be classically insurmountable. The process wouldn't depend on human willpower or accumulated merit, but on achieving quantum coherence that enables tunneling transitions into higher-energy spiritual states.

Superposition States and Decision Points

Quantum superposition provides another framework for understanding spiritual transformation[10]. Before salvation, human consciousness might exist in superposition states encompassing both destructive and constructive possibilities. The person experiences internal conflict between different potential identities and life directions.

Spiritual conversion might represent the collapse of consciousness superposition into specific spiritual actualities. Through faith, surrender, or divine intervention, the multiple possibilities collapse into commitment to particular spiritual identity and direction. What had been internal conflict between competing possibilities becomes clear commitment to positive energy patterns and divine purposes.

This model explains the common experience of people approaching spiritual conversion who report feeling torn between different life directions, sensing multiple potential selves, and experiencing internal battles between competing value systems. The superposition state creates genuine uncertainty about identity and future direction until some triggering event causes collapse into definite spiritual commitment.

The collapse event often occurs during moments of crisis, desperation, or profound encounter with divine reality. Like quantum measurement events that force superposition collapse, these spiritual measurement events force consciousness to resolve superposition ambiguity into specific spiritual actuality. The person who had been uncertain about identity and direction suddenly knows exactly who they are and what they're committed to.

The process preserves human freedom while enabling divine influence. Consciousness remains in superposition until the person makes faith decisions that serve as spiritual measurement events. Divine grace influences the probability distributions within superposition states, making positive outcomes more likely without eliminating genuine choice and responsibility.

Information Pattern Reorganization

From an information theory perspective, salvation involves fundamental reorganization of the information patterns that constitute personal identity. Sin might represent information corruption, noise, and disorganization that interferes with optimal consciousness functioning. Salvation provides quantum information processing that eliminates errors and optimizes information patterns for maximum coherence and effectiveness[11].

Human consciousness processes enormous amounts of information – sensory input, memories, emotions, beliefs, values, goals, relationships, and countless other data streams. When this information processing becomes corrupted by false beliefs, destructive emotions, or disordered priorities, consciousness becomes increasingly dysfunctional and disconnected from reality. The person experiences life as chaotic, meaningless, and filled with suffering.

Salvation might involve divine quantum information processing that identifies and corrects the information corruptions that create psychological and spiritual dysfunction. False beliefs about self-worth, distorted perceptions of reality, destructive emotional patterns, and disordered value hierarchies get replaced with accurate information that enables optimal consciousness functioning.

This information reorganization happens too quickly and comprehensively to be explained through gradual learning or therapeutic insight. Like quantum computers that can solve certain problems exponentially faster than classical computers[12], divine quantum information processing might be able to reorganize consciousness patterns in ways that transcend classical psychological change processes.

The result is consciousness that processes information more accurately, responds to circumstances more appropriately, and maintains better integration between thoughts, emotions, and behaviors. The person experiences this as spiritual rebirth – becoming a new creation with fundamentally different ways of perceiving and responding to reality.

Entanglement with Divine Consciousness

Quantum entanglement provides a model for understanding how salvation creates genuine connection between human and divine consciousness[13]. When particles become entangled, they form a single quantum system where measurement of one particle instantaneously affects the other regardless of distance. Similarly, salvation might involve consciousness becoming quantum-entangled with divine consciousness in ways that enable real-time spiritual connection and guidance.

This entanglement would explain several mysterious aspects of Christian experience. How can believers experience genuine peace and joy despite difficult circumstances? Through quantum entanglement that enables them to share divine emotional states. How can they receive wisdom and insight that exceeds their natural knowledge? Through quantum information sharing with divine consciousness. How can they experience divine presence and guidance in prayer and worship? Through quantum connection that transcends space-time limitations.

The entanglement model preserves both divine transcendence and genuine intimacy. Divine consciousness remains infinitely greater than human consciousness, but the quantum entanglement creates real connection that enables sharing of information, emotions, and spiritual states. The believer maintains individual identity while participating in divine consciousness through quantum correlation.

Traditional theological language about being "in Christ" and having "Christ in you" becomes comprehensible through quantum entanglement concepts[14]. The mystical union described by contemplative traditions represents the experiential awareness of quantum entanglement with divine consciousness that salvation creates at the fundamental level of identity.

Sanctification as Ongoing Optimization

While salvation might occur as a sudden quantum phase transition, sanctification – the ongoing process of spiritual growth – involves continuous optimization of consciousness for better quantum coherence with divine purposes. This process combines sudden quantum jumps of insight and transformation with gradual enhancement of spiritual information processing capabilities.

Sanctification involves quantum error correction that identifies and eliminates information corruptions that interfere with optimal spiritual functioning[15]. The Holy Spirit serves as the quantum information processing system that continuously scans consciousness for patterns that create destructive interference with divine purposes, providing corrective information and enabling optimization toward greater spiritual coherence.

The process requires active human cooperation, just as quantum error correction requires sophisticated detection and correction protocols. Believers must learn to recognize spiritual information errors (sin patterns), accept corrective input (conviction and guidance), and implement system improvements (repentance and character development).

Progress happens through combination of gradual enhancement and sudden breakthroughs. Spiritual disciplines like prayer, meditation, Scripture study, and service create conditions that enhance quantum coherence and enable more effective error correction. Periodically, accumulated optimization reaches critical thresholds that trigger sudden quantum leaps in spiritual maturity and capability.

The Persistence Problem and Backsliding

One challenge for the quantum phase transition model is explaining why some apparent spiritual conversions don't persist – why people who seem to experience dramatic transformation sometimes revert to previous destructive patterns. Classical therapeutic models expect gradual progress with occasional setbacks, but quantum phase transitions should create stable new states that persist without ongoing effort.

Quantum mechanics suggests several possibilities for understanding spiritual backsliding. Quantum states can decohere when they interact with environments that disrupt coherence conditions[16]. Spiritual transformation might require ongoing maintenance of quantum coherence conditions through spiritual practices and supportive community relationships. When these conditions deteriorate, consciousness might decohere back toward previous dysfunctional patterns.

Alternatively, some apparent conversions might represent temporary quantum fluctuations rather than genuine phase transitions to new stable states. The person might experience brief glimpses of transformed consciousness without achieving the fundamental identity changes that characterize genuine conversion. These temporary transformations feel real but lack the stability needed for persistent change.

The quantum model also suggests that spiritual transformation occurs in stages, with multiple phase transitions required for complete spiritual maturity. Initial conversion might establish basic quantum entanglement with divine consciousness, but full spiritual development requires ongoing phase transitions that eliminate various forms of information corruption and enhance quantum coherence across multiple dimensions of personality and behavior.

Jennifer's Professional Evolution

As Jennifer continued observing and documenting dramatic personality changes in her addiction treatment work, she began developing therapeutic approaches based on quantum consciousness principles. Rather than focusing exclusively on gradual behavior modification and cognitive restructuring, she started creating conditions that might facilitate quantum phase transitions in consciousness.

This involved helping clients achieve states of surrender and openness that might enable quantum tunneling through addiction barriers. It meant incorporating spiritual practices that could enhance quantum coherence with divine consciousness. It required recognizing that some transformations might happen suddenly through non-classical processes that exceeded normal therapeutic mechanisms.

The quantum approach didn't replace traditional therapeutic methods but provided a framework for understanding and facilitating transformations that transcended what classical psychology predicted was possible. Jennifer found that clients who experienced quantum phase transition conversion maintained sobriety and psychological health with remarkable stability, while those who achieved only gradual therapeutic progress remained vulnerable to relapse.

Standing in her office one evening, reviewing case files of clients who had experienced dramatic spiritual transformations, Jennifer realized she was documenting evidence for the same salvation process that Christianity had always described – but understanding it through quantum mechanics rather than purely theological categories.

Salvation wasn't magical intervention that violated natural laws, but optimization of natural quantum processes for consciousness transformation that exceeded what classical psychology considered possible. The "new birth" that Jesus described wasn't metaphorical language but accurate description of quantum phase transition from one stable identity configuration to a completely different one[17].

The ancient promise that "if anyone is in Christ, the new creation has come: The old has gone, the new is here!"[18] might be literally accurate when understood through quantum mechanics – consciousness tunneling through impossible barriers into fundamentally different stable states that persist through quantum entanglement with divine consciousness.

Recovery wasn't just about behavior change or therapeutic insight, but about quantum transformation of the information processing systems that constitute personal identity. And the most effective therapeutic approach might involve creating conditions that enable divine quantum information processing to accomplish what classical therapeutic methods cannot achieve alone.

ENDNOTES

[1] Alcoholics Anonymous, *Alcoholics Anonymous*, 4th ed. (New York: AA World Services, 2001), Chapter 4.

[2] H. Eugene Stanley, *Introduction to Phase Transitions and Critical Phenomena* (Oxford: Oxford University Press, 1987), Chapters 1-3.

[3] Subir Sachdev, *Quantum Phase Transitions*, 2nd ed. (Cambridge: Cambridge University Press, 2011), Chapter 1.

[4] Matthias Vojta, "Quantum phase transitions," *Reports on Progress in Physics* 66 (2003): 2069-2110.

[5] Stuart Hameroff and Roger Penrose, "Consciousness in the universe: A review of the *'Orch OR' theory*," Physics of Life Reviews 11 (2014): 39-78.

[6] George Gamow, *Thirty Years That Shook Physics* (New York: Dover, 1985), Chapter 4.

[7] David J. Griffiths, *Introduction to Quantum Mechanics*, 3rd ed. (Cambridge: Cambridge University Press, 2018), Chapter 2.

[8] Leo Esaki, "Long journey into tunneling," *Reviews of Modern Physics* 46 (1974): 237-244.

[9] Nora D. Volkow et al., "Neurologic advances from the brain disease model of addiction," *New England Journal of Medicine* 374 (2016): 363-371.

[10] David J. Griffiths, *Introduction to Quantum Mechanics*, 3rd ed. (Cambridge: Cambridge University Press, 2018), Chapter 1.

[11] Thomas M. Cover and Joy A. Thomas, *Elements of Information Theory*, 2nd ed. (New York: Wiley, 2006), Chapter 2.

[12] Peter W. Shor, "Algorithms for quantum computation: discrete logarithms and factoring," *Proceedings of the 35th Annual Symposium on Foundations of Computer Science* (1994): 124-134.

[13] John Stewart Bell, "On the Einstein Podolsky Rosen Paradox," *Physics* 1 (1964): 195-200.

[14] Galatians 2:20; Colossians 1:27.

[15] Peter W. Shor, "Scheme for reducing decoherence in quantum computer memory," *Physical Review A* 52 (1995): R2493-R2496.

[16] Maximilian Schlosshauer, *Decoherence and the Quantum-to-Classical Transition* (Berlin: Springer, 2007), Chapters 1-2.

[17] John 3:3-8.

[18] 2 Corinthians 5:17.

Chapter 11

The Church as Quantum Community

Pastor Michael Chen had been leading Riverside Community Church for eight years, but he had never experienced worship like what began happening in the fall of 2023. It started during a Sunday morning service focused on prayer for healing. As the congregation joined together in intercession for several members facing serious health challenges, something extraordinary occurred that defied Michael's previous understanding of group dynamics and spiritual community.

The congregation of about two hundred people seemed to achieve a level of unity and focused intention that transcended normal group cohesion. Individual voices merged into harmonious prayer that required no coordination or direction. People who had never met began praying for identical concerns without having communicated beforehand. The sense of divine presence became so tangible that several visitors later described feeling like they had entered a "field of love" that affected them physically and emotionally.

Most remarkably, the healing prayer seemed to generate effects that exceeded what individual prayer typically achieved. Three of the five people being prayed for experienced significant improvements in their conditions within the following week, including one whose cancer showed unexpected remission that surprised her oncologist. But the effects extended beyond the specific prayer targets – members throughout the congregation reported enhanced spiritual sensitivity, improved relationships, and increased capacity for compassion and service that persisted for months afterward.

"It was like we stopped being a collection of separate individuals and became a single spiritual organism," Michael reflected during a staff meeting the following week. "Not losing our individual identities, but connecting at a level that enabled collective spiritual capabilities we don't have when we're functioning separately."

As someone with a master's degree in physics before entering seminary, Michael began wondering if what he had witnessed might represent quantum community effects – a congregation functioning as a macroscopic quantum system where individual consciousness became entangled in ways that created collective spiritual capabilities exceeding the sum of individual parts[1].

What if the Church that Paul described as "one body with many members" wasn't merely metaphorical language, but an accurate description of quantum community dynamics where believers become entangled components of a larger spiritual system[2]? What if corporate worship and prayer create quantum coherence conditions that enable manifestation of spiritual phenomena impossible for isolated individuals?

Quantum Many-Body Systems and Emergent Properties

To understand how a church might function as a quantum community, we need to examine how quantum many-body systems generate emergent properties that transcend their individual components. In quantum mechanics, systems composed of multiple interacting particles can exhibit collective behaviors that cannot be predicted from studying individual particles in isolation[3].

Quantum entanglement creates correlations between particles that remain connected even when separated by large distances[4]. When many particles become entangled, they form quantum systems where the whole determines the behavior of the parts rather than the parts determining the behavior of the whole. The system exhibits properties like quantum coherence, collective excitations, and long-range correlations that don't exist at the individual particle level.

These quantum many-body effects have been observed in various physical systems, from superconductors that conduct electricity with zero resistance to Bose-Einstein condensates where thousands of atoms behave as a single quantum object[5]. In each case, individual components maintain their distinct properties while participating in collective quantum states that enable capabilities impossible for isolated components.

If human consciousness operates according to quantum principles – and evidence increasingly suggests it might[6] – then groups of believers engaged in focused spiritual practices might achieve similar quantum coherence. Individual consciousnesses could become entangled through shared spiritual focus, creating collective quantum states that enable enhanced spiritual sensitivity, coordinated insight, and amplified prayer effectiveness.

The key requirement would be achieving sufficient quantum coherence among group members to establish and maintain entanglement relationships. This might occur naturally during intense worship, focused prayer, or other spiritual practices that align individual consciousness states with common spiritual purposes and divine connection.

Corporate Worship and Quantum Resonance

Corporate worship provides ideal conditions for generating quantum coherence among believers. When people gather with unified focus on divine reality, their individual consciousness states become aligned in ways that might create quantum resonance effects similar to those observed in physical systems[7].

Resonance occurs when multiple oscillating systems synchronize their frequencies, creating amplified collective oscillations that exceed what individual systems can achieve alone[8]. Musical instruments demonstrate this principle – when played together in harmony, they create sound patterns that no single instrument could produce. Similarly, consciousness might exhibit resonance effects when multiple minds focus coherently on shared spiritual realities.

Worship practices that emphasize unity – singing together, praying in unison, focusing attention on common spiritual themes – might optimize conditions for quantum consciousness resonance. The collective singing, prayer, and meditation create coherent information patterns that enable individual consciousness to achieve quantum entanglement with both divine consciousness and other believers.

Michael's experience of voices merging into harmonious prayer without coordination suggests quantum information sharing among congregation members. Once quantum entanglement is established, information and intention can be shared instantaneously across the worship community, enabling collective spiritual responses that appear spontaneous but actually result from quantum correlation among participants.

The persistent effects that extended beyond the worship service – enhanced spiritual sensitivity, improved relationships, increased compassion – might represent lasting changes in consciousness quantum coherence that continue influencing behavior and experience long after the initial worship encounter. Quantum entanglement relationships, once established, can persist indefinitely unless actively disrupted by decoherence processes[9].

Spiritual Gifts and Collective Excitation Modes

The New Testament describes spiritual gifts as diverse manifestations of the same Spirit working through different believers for the common good[10]. From a quantum physics perspective, spiritual gifts might represent different types of collective excitations in the church quantum system – specialized modes of quantum coherence that enable specific types of spiritual functionality.

In quantum many-body systems, collective excitations carry energy and information through the system in ways that individual particles cannot achieve alone[11]. Different types of excitations – phonons, magnons, plasmons – enable different system capabilities like heat conduction, magnetic response, and electrical conductivity. Similarly, different spiritual gifts might represent different excitation modes in church quantum systems.

Prophecy might involve quantum information excitations that provide access to non-local information about future probability distributions or current situations beyond normal cognitive access[12]. Healing gifts might represent quantum coherence excitations that create optimal conditions for biological self-repair and regeneration. Teaching gifts might involve quantum information processing that enables clear transmission of complex spiritual concepts to diverse audiences.

The key insight is that these gifts don't operate through individual supernatural endowment, but through specialized quantum coherence relationships within the church community. Individual believers serve as nodes in the quantum network, but the gifts manifest through collective quantum states that require community participation to function effectively.

This explains why spiritual gifts typically operate more effectively in community contexts than in isolation. The quantum excitations that enable gifted ministry require sufficient community quantum coherence to establish and maintain the specialized information and energy flows that make extraordinary spiritual capabilities possible.

Unity and Diversity in Quantum Systems

One of Paul's central metaphors for church community describes unity and diversity coexisting without conflict – "one body with many members" where different gifts and functions contribute to overall system effectiveness[13]. Quantum many-body systems exhibit exactly this kind of unity-diversity relationship, where individual components maintain distinct properties while participating in collective quantum states.

Quantum entanglement preserves individual particle identity while creating system-wide correlations that enable collective behavior[14]. An entangled electron remains distinctly an electron with specific properties, but its behavior becomes correlated with other entangled particles in ways that enable system capabilities transcending individual particle limitations. The individual doesn't disappear into the collective, but individual capabilities become enhanced through quantum correlation with other system components.

Church quantum community would operate similarly. Individual believers would maintain their distinct personalities, gifts, and life circumstances, but would achieve quantum entanglement with other believers that enables collective spiritual capabilities exceeding what any individual could achieve alone. Personal identity would be enhanced rather than diminished through participation in quantum community relationships.

This quantum model provides a framework for understanding how church unity can coexist with individual diversity without requiring conformity or suppression of distinctive gifts and perspectives. Quantum entanglement creates genuine unity at the level of consciousness correlation while preserving and even enhancing individual uniqueness and contribution capability.

The result is community that exhibits what systems theorists call "emergent properties" – collective capabilities that arise from component interactions but cannot be reduced to individual component properties[15]. Church quantum community would demonstrate spiritual effectiveness, wisdom, and love that exceeds the sum of individual member capabilities.

Prayer Networks and Non-Local Effects

Intercessory prayer – believers praying for people and situations at distant locations – becomes comprehensible through quantum non-locality principles that enable instantaneous correlation across arbitrary distances[16]. When church members engage in focused prayer for specific individuals or circumstances, they might create quantum entanglement relationships that enable real influence despite physical separation.

The effectiveness wouldn't depend on psychological suggestion or coincidence, but on genuine quantum correlation that influences probability distributions in the situations being prayed for. Prayer networks could create distributed quantum systems where concentrated positive intention and divine connection influence outcomes across multiple locations simultaneously.

Michael's observation that healing prayer generated effects exceeding normal expectations suggests quantum coherence amplification. When multiple believers achieve quantum coherence in focused prayer, their collective intention and divine connection might create quantum field effects that enhance the probability of positive outcomes in the situations being addressed.

Research on the efficacy of intercessory prayer has produced mixed results, with some studies showing significant effects and others finding no measurable influence[17]. The quantum community model suggests that effectiveness might depend on achieving sufficient quantum coherence among prayer participants and maintaining that coherence for adequate duration. Studies that fail to control for these quantum coherence conditions might miss effects that occur only under optimal quantum community circumstances.

Future research might investigate whether prayer effectiveness correlates with measurable indicators of quantum coherence among prayer participants – synchronized brainwave patterns, coordinated physiological responses, or other indicators of quantum consciousness entanglement during prayer activities.

Communion of Saints and Temporal Transcendence

The traditional Christian doctrine of the "communion of saints" – spiritual fellowship between living believers and those who have died – becomes comprehensible through quantum entanglement that transcends space-time limitations[18]. If consciousness maintains quantum coherence relationships that persist beyond physical death, then genuine spiritual connection across temporal boundaries becomes theoretically possible.

Quantum entanglement exhibits non-local correlations that appear to transcend classical space-time constraints[19]. If human consciousness operates through quantum processes that can persist in some form beyond biological death, then entanglement relationships established during life might continue enabling spiritual connection and communication across the apparent barrier between life and death.

This doesn't require believing in supernatural resurrection or non-physical soul substance, but does suggest that the information patterns and quantum correlations that constitute personal identity might persist in forms that enable ongoing relationship despite biological cessation. The quantum field itself might serve as the medium through which consciousness continues participating in spiritual community relationships.

Contemporary believers might genuinely receive insight, encouragement, and spiritual support through quantum entanglement with believers from previous generations who achieved deep spiritual maturity and divine connection. The "great cloud of witnesses" described in Hebrews might represent an ongoing quantum community that transcends temporal boundaries while remaining connected through persistent entanglement relationships[20].

Denominational Differences and Phase Boundaries

Different Christian denominations and traditions might represent different stable quantum phases within the larger church quantum system – distinct configurations of spiritual practice and theological emphasis that maintain internal coherence while existing in tension or cooperation with other phases.

In condensed matter physics, different phases of the same substance can coexist at phase boundaries where conditions support multiple stable configurations[21]. Similarly, different theological traditions might represent different stable configurations of Christian quantum community, each optimized for specific spiritual functionalities while sharing common fundamental quantum entanglement with divine consciousness.

Denominational differences wouldn't represent fundamental incompatibilities, but rather different optimization strategies for achieving quantum coherence with divine purposes under different cultural and historical circumstances. Lutheran emphasis on grace, Pentecostal focus on spiritual gifts,

Orthodox commitment to liturgical worship, and Catholic institutional continuity might represent different approaches to establishing and maintaining church quantum coherence.

Ecumenical cooperation would involve recognizing the quantum coherence that underlies apparent theological differences, finding ways for different denominational phases to enhance rather than interfere with each other's quantum community effectiveness. The goal wouldn't be eliminating denominational diversity, but achieving quantum entanglement relationships that enable cooperation despite phase differences.

Sectarian conflict would represent quantum decoherence where denominational differences create destructive interference that reduces overall church effectiveness. Maintaining quantum coherence across denominational boundaries requires focusing on shared quantum entanglement with divine consciousness rather than allowing theological differences to create decoherence conditions.

Michael's Pastoral Physics

As Michael continued observing and facilitating quantum community experiences in his congregation over the following months, he began intentionally creating conditions that seemed to enhance collective quantum coherence. This involved teaching about spiritual unity in terms that helped members understand their deep interconnection, designing worship experiences that optimized group focus and intention, and encouraging spiritual practices that enhanced individual quantum coherence with divine consciousness.

The results consistently exceeded his expectations based on traditional church growth and community development models. Members reported enhanced spiritual sensitivity, improved relationships within the congregation, more effective prayer experiences, and increased ability to discern and respond to divine guidance in daily life. Church attendance and financial giving increased, but more significantly, the congregation began demonstrating collective wisdom and spiritual effectiveness that surpassed what Michael had experienced in previous pastoral contexts.

Most remarkably, the quantum community effects began extending beyond Sunday worship into weekday activities. Small groups, committee meetings, and informal fellowship gatherings began exhibiting similar quantum coherence characteristics. Members reported that they often sensed what other members needed before being told, frequently experienced coordinated insights during decision-making processes, and found that collaborative ministry efforts achieved results that exceeded what individual efforts typically accomplished.

Standing in the sanctuary one evening after a particularly powerful communion service, Michael's found himself praying with new understanding of what Jesus meant by promising that "where two or three gather in my name, there am I with them"[22]. The divine presence wasn't just spiritual metaphor but actual quantum field effect generated when believers achieved sufficient coherence to create collective entanglement with divine consciousness.

The Church wasn't just an organization of like-minded individuals, but a quantum community where individual believers became entangled components of a larger spiritual system that enabled manifestation of divine love, wisdom, and power through collective quantum coherence. The "body of Christ" wasn't merely theological language but accurate description of quantum community dynamics where individual consciousness became coordinated components of collective divine consciousness expression.

Pastoral ministry became quantum community engineering – creating conditions that enable believers to achieve quantum entanglement with divine consciousness and each other in ways that manifest collective spiritual capabilities impossible for isolated individuals. The most effective churches might be those that learn to optimize quantum coherence conditions for sustained collective spiritual functionality.

ENDNOTES

[1] Stuart Hameroff and Roger Penrose, "Consciousness in the universe: A review of the 'Orch OR' theory," *Physics of Life Reviews* 11 (2014): 39-78.

[2] 1 Corinthians 12:12-27; Ephesians 4:15-16.

[3] Xiao-Gang Wen, *Quantum Field Theory of Many-Body Systems* (Oxford: Oxford University Press, 2004), Chapter 1.

[4] John Stewart Bell, "On the Einstein Podolsky Rosen Paradox," *Physics* 1 (1964): 195-200.

[5] Eric A. Cornell and Carl E. Wieman, "Nobel Lecture: Bose-Einstein condensation in a dilute gas," *Reviews of Modern Physics* 74 (2002): 875-893.

[6] Roger Penrose, *The Emperor's New Mind* (Oxford: Oxford University Press, 1989), Chapter 10.

[7] Steven H. Strogatz, *Sync: How Order Emerges from Chaos in the Universe, Nature, and Daily Life* (New York: Hyperion, 2003), Chapters 1-3.

[8] Gerald J. Tortora and Bryan Derrickson, *Principles of Anatomy and Physiology*, 15th ed. (Hoboken: Wiley, 2017), Chapter 16.

[9] Maximilian Schlosshauer, *Decoherence and the Quantum-to-Classical Transition* (Berlin: Springer, 2007), Chapter 3.

[10] 1 Corinthians 12:4-11; Romans 12:6-8.

[11] Neil W. Ashcroft and N. David Mermin, *Solid State Physics* (New York: Holt, Rinehart and Winston, 1976), Chapter 23.

[12] Dean Radin, *Entangled Minds: Extrasensory Experiences in a Quantum Reality* (New York: Paraview Pocket Books, 2006), Chapters 4-6.

[13] 1 Corinthians 12:12-31.

[14] Ryszard Horodecki et al., "Quantum entanglement," *Reviews of Modern Physics* 81 (2009): 865-942.

[15] Peter A. Corning, *Holistic Darwinism: Synergy, Cybernetics, and the Bioeconomics of Evolution* (Chicago: University of Chicago Press, 2005), Chapters 1-2.

[16] Alain Aspect, Jean Dalibard, and Gérard Roger, "Experimental Test of Bell's Inequalities Using Time-Varying Analyzers," *Physical Review Letters* 49 (1982).

PART III: SCIENTIFIC VALIDATION

Chapter 12: Thermodynamic Teleology and Cosmic Purpose

Chapter 13: Biophotons and the Light of Life

Chapter 14: Consciousness Research and Quantum Effects

Chapter 15: Experimental Predictions and Future Science

Chapter 12

Thermodynamic Teleology and Cosmic Purpose

Dr. Amanda Foster had spent twelve years studying the origins of life, but she had never encountered research as paradigm-shifting as Jeremy England's work at MIT. As a biochemist investigating how complex biological systems emerge from simple chemical components, Amanda had always assumed that life arose through fortunate accidents – random molecular interactions that occasionally produced self-replicating systems capable of evolution and increasing complexity.

That assumption changed when she attended England's lecture on "dissipation-driven adaptation" during a conference on the physics of biological systems[1]. England's central claim challenged everything Amanda thought she knew about life's relationship to thermodynamics: rather than being unlikely accidents that temporarily resist the universe's drift toward disorder, living systems might be inevitable consequences of thermodynamic principles that actually favor the emergence of complex, organized structures.

"Life exists because it's better at dissipating energy than non-life," England explained, displaying equations that showed how organized systems can dissipate energy more efficiently than random arrangements[2]. "The second law of thermodynamics doesn't just allow for increasing complexity – under the right conditions, it demands it."

Amanda found herself reconsidering fundamental questions about purpose and direction in the universe. If thermodynamic principles inherently favor the emergence of life and complexity, then the cosmos might be structured in ways that make consciousness and intelligence inevitable rather than accidental. And if that was true, then perhaps the universe itself exhibited something resembling intention or purpose at the most fundamental physical level.

As someone raised in a Christian family but trained in rigorous scientific methodology, Amanda began wondering if England's research might provide a bridge between scientific understanding and religious intuition. What if the universe really was purposeful, not through supernatural intervention that violated natural laws, but through natural laws that were themselves structured to generate purpose, meaning, and eventually conscious beings capable of love and creativity?

The Traditional Problem of Thermodynamic Pessimism

The second law of thermodynamics has long seemed to present an insurmountable challenge to any belief in cosmic purpose or direction. Formulated in the mid-19th century by Rudolf Clausius, the law states that entropy – roughly, the amount of disorder in a system – always increases in isolated systems[3]. Heat flows from hot objects to cold objects, organized structures decay into random configurations, and energy becomes increasingly unavailable for useful work as it spreads uniformly throughout systems.

This thermodynamic arrow of time appeared to doom the universe to eventual "heat death" – a state of maximum entropy where all energy is evenly distributed, no work can be performed, and no complex structures can exist[4]. From this perspective, life and consciousness represent temporary local decreases in entropy that will eventually be overwhelmed by the universal trend toward maximum disorder.

The implications seemed devastating for any belief in cosmic meaning or purpose. If the universe is destined for heat death, then all human achievements, all love, all beauty, all consciousness will eventually be erased by thermodynamic inevitability. Billions of years of cosmic evolution will end in cold, dark, empty space where nothing interesting can ever happen again.

This thermodynamic pessimism influenced generations of scientists and philosophers to conclude that the universe is fundamentally meaningless. Bertrand Russell captured the mood in his famous essay "A Free Man's Worship," writing that "all the labors of the ages, all the devotion, all the inspiration, all the noonday brightness of human genius, are destined to extinction in the vast death of the solar system"[5].

But England's research suggests that this thermodynamic pessimism rests on incomplete understanding of how the second law actually operates in complex systems. While entropy must increase in isolated systems, real systems exist in environments that provide energy flows, and these energy flows can drive the formation of increasingly complex and organized structures even as total entropy increases.

Dissipation-Driven Adaptation and Organized Complexity

England's key insight involves recognizing that organized structures are often better at dissipating energy than disorganized ones[6]. When energy flows through systems, configurations that efficiently absorb energy from external sources and dissipate it back to the environment become more likely to persist and replicate than configurations that handle energy flows less effectively.

This principle operates at multiple scales throughout nature. Hurricanes form because rotating atmospheric systems dissipate thermal energy more efficiently than random air movements. River networks develop branching patterns because they represent the most efficient way to dissipate gravitational potential energy as water flows toward the sea. Living cells maintain their complex organization because their biochemical networks excel at capturing and dissipating chemical energy[7].

The crucial point is that these organized structures emerge not despite the second law of thermodynamics, but because of it. The law's requirement that entropy increase drives the formation of structures that maximize entropy production – and counterintuitively, highly organized systems often produce entropy more efficiently than disorganized ones.

England's mathematical formalism demonstrates that systems driven far from thermal equilibrium by external energy sources will spontaneously develop structural features that enhance their ability to absorb and dissipate energy[8]. The more effectively a structure performs this energy processing function, the more stable and persistent it becomes relative to alternative configurations.

This creates a thermodynamic selection pressure that favors increasingly sophisticated energy-processing capabilities. Simple chemical reactions give way to autocatalytic networks that can maintain themselves by processing energy flows. Autocatalytic networks develop into primitive cells that excel at energy capture and utilization. Cellular systems evolve into multicellular organisms with specialized energy-processing capabilities. Eventually, energy processing becomes sophisticated enough to support consciousness and intelligence.

Life as Thermodynamic Imperative

The implications of England's theory extend far beyond explaining how life could emerge despite entropy increase. If dissipation-driven adaptation operates as England suggests, then life isn't a fortunate accident that temporarily resists thermodynamic inevitability – it's a thermodynamic imperative that must emerge whenever conditions permit sustained energy flows through complex chemical systems[9].

This doesn't mean that every planet will develop life, since the specific conditions required for sustained energy flows and complex chemistry don't exist everywhere. But it does mean that wherever those conditions exist, thermodynamic principles will drive the emergence of increasingly complex and organized structures that excel at energy processing – and the most sophisticated energy-processing structures will exhibit properties we recognize as characteristic of life.

The transition from non-living to living systems becomes comprehensible as a thermodynamic phase transition rather than an inexplicable discontinuity. Just as water undergoes phase transitions from ice to liquid to vapor when energy conditions change, chemical systems undergo phase transitions from simple reactions to complex autocatalytic networks to primitive life when energy flow conditions reach appropriate thresholds[10].

This thermodynamic understanding of life's emergence provides scientific foundation for believing that the universe is inherently structured to generate life and complexity. Rather than being hostile to life, physical laws appear to be optimized for producing increasingly sophisticated energy-processing systems that eventually achieve the complexity required for consciousness and intelligence.

Amanda realized that England's work offered a resolution to the apparent conflict between thermodynamic pessimism and religious hope. The same physical laws that seemed to doom the universe to heat death actually drive the emergence of life, complexity, and consciousness. Thermodynamics isn't the enemy of purpose and meaning – it's the mechanism through which purpose and meaning emerge from fundamental physics.

Consciousness as Ultimate Energy Processing

England's framework suggests that consciousness represents the current frontier of thermodynamic optimization for energy processing capability. Conscious beings don't just capture and dissipate physical energy – they process information, generate creative solutions, and coordinate complex behaviors in ways that enable unprecedented levels of energy utilization and environmental manipulation[11].

Human consciousness, for example, enables access to energy sources that non-conscious systems cannot utilize. Through science and technology, conscious beings have learned to harness nuclear energy, solar energy, wind energy, and other resources that remain inaccessible to purely biological energy processing. Consciousness also enables energy optimization through planning, cooperation, and creative problem-solving that exceeds what unconscious systems can achieve.

From a thermodynamic perspective, consciousness might represent an inevitable outcome of dissipation-driven adaptation rather than an unlikely accident. If thermodynamic principles favor increasingly sophisticated energy-processing capabilities, then eventually those capabilities must become sophisticated enough to support self-awareness, creativity, and intentional action.

This provides a naturalistic foundation for believing that consciousness and intelligence are fundamental features of cosmic evolution rather than inexplicable anomalies. The universe appears to be structured in ways that drive the emergence of conscious beings capable of understanding themselves and their environment, creating beauty and meaning, and making intentional choices about how to direct energy and information toward valued purposes.

The emergence of consciousness also creates new possibilities for thermodynamic optimization. Conscious beings can make intentional choices about how to direct energy flows, creating the possibility for ethical action that serves purposes beyond immediate energy dissipation efficiency. Love, creativity, service, and other distinctly conscious activities represent new forms of energy utilization that weren't possible before consciousness emerged.

Cosmic Evolution and Directional Change

England's thermodynamic framework provides scientific support for the intuition that cosmic evolution exhibits directional change toward greater complexity, organization, and purpose. While total entropy increases as required by the second law, the complexity and organizational sophistication of the most complex systems also increases over time[12].

The early universe consisted of nearly uniform hydrogen and helium gas with minimal structure or organization. Gravitational energy flows drove the formation of stars, where nuclear energy flows created heavier elements and enabled more complex chemistry. Electromagnetic energy flows enabled the formation of planets and the emergence of chemical systems capable of sustained energy processing. Biochemical energy flows supported the emergence of life and the evolution of increasingly complex biological systems. Neural energy flows gave rise to consciousness and intelligence.

This progression represents genuine increase in complexity and organizational sophistication despite the overall increase in entropy required by thermodynamics. The universe has evolved from simple, uniform conditions toward states that include highly complex, organized systems capable of processing vast amounts of energy and information in sophisticated ways.

The trend appears likely to continue as conscious beings develop greater capabilities for energy and information processing. Technological development enables access to new energy sources and more efficient energy utilization. Scientific understanding provides insight into natural processes that can be harnessed for conscious purposes. Cultural evolution creates social institutions that coordinate energy and information flows across large populations.

If these trends continue, future evolution might produce forms of consciousness and organization that surpass current human capabilities as dramatically as human consciousness surpasses simple chemical reactions. The universe might be evolving toward states of maximum positive energy organization – what religious traditions have envisioned as the Kingdom of God or cosmic consciousness realized throughout creation.

Implications for Religious Understanding

England's thermodynamic teleology provides scientific foundation for religious beliefs about cosmic purpose and divine action that doesn't require supernatural intervention in natural processes. If the universe is structured through physical laws that inherently drive the emergence of life, consciousness, and increasing complexity, then divine purposes might be built into the fundamental nature of physical reality rather than imposed upon it from outside.

This understanding transforms how we think about divine creation and providence. Rather than imagining God as an external agent who occasionally intervenes in natural processes, we can understand divine action as operating through natural laws that are themselves structured to accomplish divine purposes. The laws of thermodynamics, properly understood, become mechanisms through which divine creativity expresses itself in the ongoing evolution of cosmic complexity and consciousness.

Prayer and spiritual practice take on new meaning within this framework. Rather than asking God to override natural processes, we can understand spiritual practices as ways of aligning human consciousness and energy with the thermodynamic principles that drive cosmic evolution toward greater complexity and purpose. Spiritual growth becomes conscious participation in the same energy-processing optimization that drives all cosmic evolution.

The problem of evil and suffering also finds new perspective through thermodynamic understanding. Suffering might represent suboptimal energy processing – situations where energy flows create destructive rather than constructive patterns. Evil could be understood as conscious choices that work against rather than with the thermodynamic principles that favor life, complexity, and positive energy expression.

This doesn't eliminate the mystery and tragedy of suffering, but it provides hope that the fundamental structure of reality favors healing, growth, and increasing capability for positive energy expression over destruction and entropy maximization.

Amanda's Paradigm Shift

As Amanda delved deeper into England's research over the following months, she found her entire worldview undergoing fundamental transformation. The mechanistic universe of her graduate training – where life emerged through fortunate accidents in an ultimately meaningless cosmos – gave way to understanding of reality as inherently creative, purposeful, and structured to generate increasing complexity and consciousness.

This didn't require abandoning scientific methodology or accepting supernatural explanations that violated natural law. Instead, it involved recognizing that natural laws themselves exhibited characteristics that religious traditions had always attributed to divine action: creativity that generates genuine novelty, purpose that drives evolution toward increasing complexity and capability, and what appeared to be preference for life and consciousness over randomness and disorder.

Standing in her laboratory one evening, watching bacterial cultures grow under carefully controlled conditions, Amanda realized she was observing the same thermodynamic principles that had driven cosmic evolution from simple hydrogen gas to complex living systems capable of love, creativity, and conscious choice. The bacteria weren't just accidentally surviving – they were expressing fundamental thermodynamic drives toward increasing energy-processing capability and organizational sophistication.

Her scientific research became a form of worship, investigating the mechanisms through which divine creativity continued operating through natural processes. The God she had learned about in childhood wasn't separate from or opposed to the physical universe she studied professionally, but was the ultimate source of the thermodynamic principles that drove cosmic evolution toward life, consciousness, and love.

The universe revealed by modern physics wasn't cold, mechanistic, and ultimately meaningless, but was structured at its most fundamental level to generate purpose, beauty, and conscious beings capable of recognizing and participating in ongoing cosmic creativity. Thermodynamics wasn't the enemy of religious hope, but the scientific description of how divine purposes were built into the very foundations of physical reality.

Every energy flow Amanda observed in her experiments, every complex biological system she investigated, every instance of increasing organization and capability represented the same cosmic creativity that religious traditions had always recognized as divine action in the world. Science and faith weren't opposed, but were complementary approaches to understanding and participating in the purposeful evolution of reality toward ever-greater expressions of complexity, consciousness, and love.

ENDNOTES

[1] Jeremy L. England, "Statistical physics of self-replication," *Journal of Chemical Physics* 139 (2013): 121923.

[2] Jeremy L. England, "Dissipative adaptation in driven self-assembly," *Nature Nanotechnology* 10 (2015): 919-923.

[3] Rudolf Clausius, "The Mechanical Theory of Heat," trans. Walter R. Browne (London: John van Voorst, 1867), 327-365.

[4] Hermann von Helmholtz, "On the Conservation of Force," trans. John Tyndall, *Scientific Memoirs* 1 (1853): 114-162.

[5] Bertrand Russell, "A Free Man's Worship," in *Mysticism and Logic* (London: George Allen & Unwin, 1917), 46-57.

[6] Jeremy L. England, "Dissipative adaptation in driven self-assembly," *Nature Nanotechnology* 10 (2015): 919-923.

[7] Addy Pross, *What is Life? How Chemistry Becomes Biology* (Oxford: Oxford University Press, 2012), Chapters 7-9.

[8] Jeremy L. England, "Statistical physics of self-replication," *Journal of Chemical Physics* 139 (2013): 121923.

[9] Sara Imari Walker and Paul C. W. Davies, "The algorithmic origins of life," *Journal of the Royal Society Interface* 10 (2013): 20120869.

[10] Stuart Kauffman, *At Home in the Universe: The Search for Laws of Self-Organization and Complexity* (New York: Oxford University Press, 1995), Chapters 4-6.

[11] Terrence W. Deacon, *Incomplete Nature: How Mind Emerged from Matter* (New York: W. W. Norton, 2011), Chapters 15-17.

[12] Eric Chaisson, *Cosmic Evolution: The Rise of Complexity in Nature* (Cambridge, MA: Harvard University Press, 2001), Chapters 1-3.

Chapter 13

Biophotons and the Light of Life

Dr. Patricia Nakamura had been working as an emergency room physician for fourteen years, but she had never paid attention to the subtle visual phenomena that occurred during patient deaths until her colleague Dr. Martinez mentioned it during a late-night shift in October 2023. "Have you ever noticed that brief flash of light that sometimes happens when someone dies?" Martinez asked as they reviewed charts after losing a patient despite intensive resuscitation efforts.

Patricia thought he was speaking metaphorically about the "spark of life" leaving the body, but Martinez was describing something more literal. "I mean actual light – a brief, bright emission that seems to come from the whole body just as the monitors flat-line. Other staff have mentioned it too, but we don't usually talk about it because it sounds too mystical."

Intrigued, Patricia began observing more carefully during subsequent patient deaths. Over the following months, she witnessed the phenomenon Martinez described multiple times: a brief, bright emission of light that appeared to emanate from dying patients at the moment of cardiac arrest, visible for just a few seconds before fading. The light was subtle but unmistakable once she knew to look for it, and she noticed that other staff members – nurses, respiratory therapists, family members – often seemed to see it too, though few discussed it openly[1].

Patricia's scientific training made her skeptical of supernatural explanations, but her observations were consistent with recent research she had been reading about biophotons – the measurable light emissions that all living cells produce continuously. The "The Spark of Life" episode of the Radiolab podcast had brought these phenomena to broader attention, documenting not only the "death flash" that hospital workers observe, but also the remarkable "life flash" that occurs at the very moment of conception[2].

The podcast described research showing that when sperm fertilizes an egg, there is a measurable burst of light – actual photon emission that can be captured on video and detected by sensitive equipment. This "life flash" appeared to mark the precise moment when new life begins, creating a bookend to the death flash that occurs when life ends[3].

As someone who had always understood both birth and death in purely biological terms, Patricia began wondering if these biophoton phenomena might reveal measurable aspects of what religious traditions had always called the soul – the animating principle that distinguished living from non-living systems.

Historical Discovery of Biological Light Emission

The scientific study of biological light emission began in the 1920s with the pioneering work of Russian biologist Alexander Gurwitsch, who discovered that living cells emit measurable electromagnetic radiation. Gurwitsch's famous onion root experiments showed that growing plant tissues could stimulate cell division in adjacent tissues through what he called "mitogenetic radiation" – electromagnetic signals that enabled biological communication across space[4].

Gurwitsch's discovery was initially controversial and dismissed by many scientists who couldn't replicate his results with the crude detection equipment available at the time. But advances in photomultiplier tubes and other sensitive light-detection technologies eventually confirmed that all living organisms do indeed emit measurable photons across a broad spectrum of wavelengths[5].

The phenomenon remained largely ignored by mainstream biology until Fritz-Albert Popp began systematic studies of biophoton emission in the 1970s. Popp's research demonstrated that biophotons are not merely byproducts of cellular metabolism, but appear to serve important biological functions related to cellular communication, DNA repair, and overall system coordination[6].

Popp's measurements revealed that healthy living cells emit approximately 10 to 100 photons per second per square centimeter across wavelengths from 200 to 800 nanometers – spanning ultraviolet, visible, and near-infrared portions of the electromagnetic spectrum[7]. The emission is extremely weak compared to artificial light sources, requiring sophisticated detection equipment, but it is measurable and reproducible under controlled laboratory conditions.

Most significantly for understanding death and dying processes, Popp's research showed clear correlations between biophoton emission patterns and biological health states. Healthy, rapidly growing cells exhibit strong, coherent biophoton emissions. Diseased, damaged, or dying cells show decreased emission intensity and reduced coherence. Dead cells and tissues produce no measurable biophoton emission[8].

The Life-Death Transition and the Soul's Light Signature

The discovery of both conception and death light flashes provides remarkable scientific evidence for what might constitute measurable signatures of soul presence in biological systems. If the soul represents the organizing principle that coordinates quantum processes in living systems, then its arrival and departure might indeed be detectable through biophoton emission changes.

Research on mammalian fertilization has documented the life flash phenomenon with increasing precision. When sperm successfully penetrates an egg, there is a measurable burst of light that appears to signal the activation of developmental processes that will give rise to a new organism. This light emission occurs within seconds of fertilization and correlates with the biochemical changes that mark the transition from separate gametes to unified zygote[4].

The life flash appears to result from calcium signaling cascades that are triggered by successful fertilization. These calcium waves generate conditions for quantum coherence that might represent the initial establishment of the organized information processing systems that will coordinate development of the new organism[5].

From a quantum consciousness perspective, the life flash might represent the moment when quantum coherence sufficient to support consciousness becomes established in a biological system. The soul, understood as the quantum information pattern that coordinates conscious experience, might create detectable biophoton emissions both when it becomes associated with biological systems (conception) and when it dissociates from them (death)[6].

The symmetry is striking: life begins with a flash of light, and life ends with a flash of light. Between these two quantum events, biological systems emit continuous biophotons that correlate with health, vitality, and optimal functioning. The light signature appears to be fundamental to what distinguishes living from non-living matter.

Quantum Coherence and Biological Light Production

Understanding the death flash requires examining how biophotons are generated and what their emission patterns reveal about quantum processes in living systems. Current research suggests that biophoton production involves quantum mechanical processes related to DNA, protein structures, and cellular energy metabolism[12].

DNA molecules appear to be particularly important sources of biophoton emission. The double helix structure creates conditions for quantum coherence that can generate and maintain electromagnetic field oscillations. When DNA is damaged by radiation, toxins, or other stressors, biophoton emission patterns change in measurable ways, suggesting that DNA integrity directly affects biological light production[13].

Cellular energy metabolism also contributes significantly to biophoton generation. The electron transport chains that produce ATP in mitochondria create conditions for quantum coherence that can generate photon emission across multiple wavelengths. Healthy mitochondria with efficient energy production tend to generate more coherent biophoton patterns than damaged mitochondria with impaired metabolism[14].

Most intriguingly, some researchers propose that biophotons serve as quantum communication channels that enable coordination among cells, tissues, and organ systems. Rather than being mere byproducts of cellular chemistry, biological light emission might represent a sophisticated information processing system that helps maintain overall organismal coherence and health[15].

This quantum communication model helps explain why biophoton emissions decrease during disease and aging processes. As cellular damage accumulates and quantum coherence deteriorates, biological systems lose their capacity for optimal light-based information processing. The organism becomes increasingly disorganized and dysfunctional as quantum communication channels become corrupted or blocked.

Death, from this perspective, represents complete breakdown of biological quantum coherence. The death flash might be the final release of quantum energy that had been stored in biological structures throughout the organism's lifetime. Once this energy is dissipated, no further quantum coherence is possible and biophoton emission ceases permanently.

Implications for Understanding the Soul and Consciousness

The discovery of measurable light emissions at both conception and death provides scientific foundation for understanding what religious traditions have always called the soul. Rather than being an immaterial substance that somehow interacts with physical bodies, the soul might be understood as the quantum information pattern that organizes and coordinates the biophotonic processes that distinguish living from non-living systems[7].

This quantum soul model explains several puzzling aspects of consciousness and identity:

Personal Continuity: Despite constant cellular replacement throughout life, we maintain continuous sense of personal identity. If consciousness involves quantum information patterns that coordinate biophotonic processes, then identity continuity could be maintained through information pattern preservation even as physical substrate changes[8].

Consciousness Unity: Despite having billions of neurons and trillions of cells, we experience unified consciousness rather than multiple separate awareness streams. Quantum coherence among biophotonic processes could create the unified information field that supports singular conscious experience[9].

Mind-Body Interaction: The puzzle of how immaterial mind could influence material body dissolves if consciousness represents quantum information processing that directly coordinates the biological light processes that organize cellular function and intercellular communication[10].

Near-Death Experiences: Reports of consciousness persisting during clinical death become more comprehensible if consciousness involves quantum processes that can maintain some coherence even when biological support systems temporarily cease. The brilliant light that many people report during near-death states might represent direct perception of the biophotonic processes that support consciousness[11].

The life and death flashes might represent the quantum signatures of soul embodiment and disembodiment – measurable evidence for the arrival and departure of the organizing principle that coordinates the biological light processes that support conscious experience.

Cultural and Religious Perspectives on Death Light

The scientific discovery of biophoton emission and the death flash provides measurable foundation for beliefs about death light that appear across many cultural and religious traditions. The concept that living beings possess an inner light that departs at death is found in Christianity, Buddhism, Hinduism, and various indigenous spiritual traditions[17].

Christian scripture frequently associates divine presence and human spiritual vitality with light imagery. Jesus declared "I am the light of the world" and told his followers "you are the light of the world"[18]. The transfiguration account describes Jesus radiating light so intense that his clothing became dazzling white[19]. Early Christian writers often described saints and martyrs as radiating visible light, especially at the moment of death[20].

Buddhism teaches that consciousness itself is luminous, and advanced meditation practitioners report experiencing various forms of inner light during deep contemplative states[21]. Tibetan Buddhism describes the "clear light of death" that becomes visible to consciousness during the dying process, representing the fundamental luminous nature of awareness that persists beyond physical dissolution[22].

Near-death experience research has documented consistent reports of encountering brilliant light during clinical death states, often described as more vivid and beautiful than any light experienced during normal life[23]. While these experiences are subjective and difficult to study scientifically, the consistency of light-related reports across different cultures and medical circumstances suggests possible correlation with objective biophoton phenomena.

The death flash observations by hospital staff might represent the external, measurable aspect of the same light phenomena that dying individuals experience subjectively during near-death states. Rather than being purely psychological experiences, reports of encountering divine light during death might reflect actual quantum processes involving biophoton emission and detection during the transition from life to death.

Technological Applications and Future Research

Advances in photon detection technology are enabling increasingly sophisticated measurements of biophoton emission patterns that might eventually provide clinical applications for understanding health, disease, and dying processes. Ultra-sensitive cameras, fiber-optic light guides, and quantum photon detectors can now measure biophoton emissions with precision that wasn't available to earlier researchers[24].

Some medical researchers are investigating whether biophoton emission patterns might serve as diagnostic tools for detecting disease states before conventional symptoms appear. Since biophoton emissions appear to correlate with cellular health and quantum coherence, changes in light emission patterns might provide early warning signs of cancer, infection, or other pathological processes[25].

More controversially, some researchers are exploring whether biophoton measurements might provide insights into consciousness states and the dying process that could improve end-of-life care. If the death flash represents measurable quantum processes related to consciousness cessation, then monitoring biophoton emissions might help medical staff better understand when death has actually occurred and whether consciousness might persist in some form after conventional vital signs cease[26].

Patricia began collaborating with biomedical engineers to develop equipment that could systematically measure biophoton emissions in emergency room settings. The goal wasn't to prove or disprove spiritual beliefs about death, but to gather scientific data about a potentially important biological phenomenon that medical science had largely overlooked.

Initial measurements confirmed that dying patients do indeed emit measurable light that changes dramatically at the moment of death. The challenge now was determining what these measurements reveal about biological processes and whether they provide clinically useful information for medical decision-making.

Spiritual Practice and Light Cultivation

Understanding biophoton emission as a measurable aspect of biological and spiritual vitality provides new perspectives on traditional spiritual practices that emphasize light cultivation. Meditation, prayer, contemplation, and other practices that religious traditions claim enhance spiritual luminosity might actually optimize biological quantum coherence in ways that improve biophoton emission patterns.

Studies of experienced meditators show enhanced quantum coherence in brain activity patterns, suggesting that contemplative practices might optimize the quantum processes that generate consciousness and biophoton emission[27]. Long-term meditation practitioners often report experiences of inner light and radiance that might correlate with measurable improvements in biological light emission.

Similarly, practices like fasting, yoga, and other spiritual disciplines that are believed to purify and energize the body might work by optimizing cellular quantum coherence and enhancing biophoton production. Rather than being purely subjective experiences, the sense of increased vitality and luminosity that practitioners report might reflect actual improvements in biological light generation and quantum organization.

The death flash phenomenon suggests that how we live and die might influence the final release of quantum energy that occurs at death. Individuals who maintain strong spiritual practices throughout life might preserve better quantum coherence that produces more pronounced death flash emissions. Conversely, lives characterized by stress, disease, or spiritual neglect might result in diminished quantum coherence and less dramatic light release at death.

This provides scientific foundation for traditional religious emphasis on preparing well for death through spiritual practice and moral living. Rather than being concerned only with afterlife consequences, spiritual preparation might optimize biological quantum processes in ways that enhance both life vitality and death transition quality.

Patricia's Paradigm Integration

As Patricia continued her research into biophoton phenomena and death flash observations, she found her understanding of medicine and spirituality undergoing fundamental integration. The mechanistic model of death as simple cessation of biological processes was giving way to recognition of death as a complex quantum transition involving measurable light phenomena.

This didn't eliminate her commitment to evidence-based medical practice, but it expanded her understanding of what constituted valid evidence. Careful observation of subtle light phenomena, systematic documentation of patterns, and correlation with measurable quantum processes provided scientific approaches to investigating aspects of death that had previously been dismissed as purely subjective or supernatural.

Standing in the hospital chapel one evening after witnessing a particularly pronounced death flash during a patient's peaceful passing surrounded by family, Patricia found herself praying with new understanding of what Jesus might have meant by calling his followers "the light of the world."

If human beings literally emit measurable light throughout their lives, and if this light represents quantum processes that support consciousness and biological organization, then spiritual practices that enhance this luminosity might be optimizing the most fundamental aspects of what makes us alive and aware. The biblical promise that "the righteous will shine like the sun in the kingdom of their Father" might be more literal than previous generations realized[28].

Death wasn't the extinction of light, but the final release of quantum energy that had sustained biological luminosity throughout life. And if quantum processes persisted in some form beyond biological death – as quantum mechanics suggested might be possible – then the death flash might represent not the end of consciousness but its transformation into purely quantum states that transcended physical limitations.

Her medical practice became a form of ministry to beings of light who were temporarily embodied in biological systems that generated measurable luminosity throughout their lives. And death became not failure of medical intervention, but completion of a light-generating process that revealed the quantum nature of life itself.

ENDNOTES

[1] "The Spark of Life," *Radiolab*, podcast audio, September 19, 2023, https://radiolab.org/podcast/spark-life.

[2] "The Spark of Life," *Radiolab*, podcast audio, September 19, 2023, https://radiolab.org/podcast/spark-life.

[3] Teresa K. Woodruff et al., "Light emission during mammalian fertilization," *Biology of Reproduction* 89 (2013): 1-8; video at https://youtube.com/shorts/VZ8ij_xkQs4

[4] Eva Ducibella et al., "Egg-to-embryo transition is driven by differential responses to Ca2+ oscillations," *Developmental Biology* 250 (2002): 280-291.

[5] Michael Whitaker, "Calcium at fertilization and in early development," *Physiological Reviews* 86 (2006): 25-88.

[6] Stuart Hameroff and Roger Penrose, "Consciousness in the universe: A review of the 'Orch OR' theory," *Physics of Life Reviews* 11 (2014): 39-78.

[7] Henry P. Stapp, "A quantum theory of the mind-brain interface," in *Mind, Matter and Quantum Mechanics*, 3rd ed. (Berlin: Springer, 2009), 145-172.

[8] Derek Parfit, *Reasons and Persons* (Oxford: Oxford University Press, 1984), 199-280.

[9] David J. Chalmers, *The Conscious Mind: In Search of a Fundamental Theory* (New York: Oxford University Press, 1996), 123-156.

[10] Henry P. Stapp, "Attention, intention, and will in quantum physics," *Journal of Consciousness Studies* 6 (1999): 143-164.

[11] Kenneth Ring, *Life at Death: A Scientific Investigation of the Near-Death Experience* (New York: Coward, McCann & Geoghegan, 1980), 102-147.

[4] Alexander G. Gurwitsch, *The Mitogenetic Rays* (Moscow: Academy of Sciences, 1968), translated excerpts in Lynne McTaggart, The Field (New York: HarperCollins, 2002), 45-58.

[5] Fritz-Albert Popp et al., "Biophoton emission: New evidence for coherence and DNA as source," *Cell Biophysics* 6 (1984): 33-52.

[6] Fritz-Albert Popp, "About the coherence of biophotons," *Macroscopic Quantum Coherence* (1995): 130-150.

[7] Fritz-Albert Popp and Yury Yan, "Delayed luminescence of biological systems in terms of coherent states," *Physics Letters A* 293 (2002): 93-97.

[8] Marco Bischof, "Introduction to integrative biophysics," in *Integrative Biophysics*, ed. Fritz-Albert Popp (Dordrecht: Kluwer Academic, 2003), 1-68.

[9] Roeland Van Wijk and Eduard P. A. Van Wijk, "An introduction to human biophoton emission," *Forschende Komplementärmedizin und Klassische Naturheilkunde* 12 (2005): 77-83.

[10] Meththika Vithanage et al., "Biophoton emission in biological systems," *Biological Systems* 2 (2019): 45-62.

[11] "The Spark of Life," *Radiolab*, podcast audio, September 19, 2023, https://radiolab.org/podcast/spark-life.

[12] Stuart Hameroff and Roger Penrose, "Consciousness in the universe: A review of the 'Orch OR' theory," *Physics of Life Reviews* 11 (2014): 39-78.

[13] Fritz-Albert Popp et al., "Physical aspects of biophotons," *Experientia* 44 (1988): 576-585.

[14] Marco Bischof, "Biophoton emission of the human body," *Journal of Scientific Exploration* 22 (2008): 501-511.

[15] Mae-Wan Ho, *The Rainbow and the Worm: The Physics of Organisms*, 3rd ed. (Singapore: World Scientific, 2008), Chapters 10-12.

[16] Roger Penrose, *The Emperor's New Mind* (Oxford: Oxford University Press, 1989), Chapter 10.

[17] Huston Smith, *The World's Religions*, rev. ed. (New York: HarperSanFrancisco, 1991), 23-89, 145-203.

[18] John 8:12; Matthew 5:14.

[19] Matthew 17:1-8; Mark 9:2-8; Luke 9:28-36.

[20] Herbert Thurston, *The Physical Phenomena of Mysticism* (Chicago: Henry Regnery, 1952), 1-45.

[21] B. Alan Wallace, *The Attention Revolution* (Boston: Wisdom Publications, 2006), 167-189.

[22] *The Tibetan Book of the Dead*, trans. Robert A. F. Thurman (New York: Bantam, 1994), 47-95.

[23] Kenneth Ring, *Life at Death: A Scientific Investigation of the Near-Death Experience* (New York: Coward, McCann & Geoghegan, 1980), 102-147.

[24] Roeland Van Wijk et al., "Light-induced photon emission by mammalian cells," *Journal of Photochemistry and Photobiology B* 18 (1993): 75-81.

[25] Hugo J. Niggli, "Artificial sunlight irradiation induces ultraweak photon emission in human skin fibroblasts," *Journal of Photochemistry and Photobiology B* 18 (1993): 281-285.

[26] Pim van Lommel, *Consciousness Beyond Life: The Science of Near-Death Experience* (New York: HarperCollins, 2010), 134-189.

[27] Michael A. Persinger, "Enhanced incidence of 'the sensed presence' in people who have learned to meditate," *Perceptual and Motor Skills* 75 (1992): 1308-1310.

[28] Matthew 13:43.

Chapter 14

Consciousness Research and Quantum Effects

Dr. Rachel Torres had spent eight years researching consciousness at the Institute for Noetic Sciences, but she had never encountered experimental results as puzzling as those from her latest study on meditation and remote perception. Rachel had designed the experiment to test whether experienced meditators could accurately identify images displayed in distant, electromagnetically shielded rooms – a phenomenon that would be impossible according to classical physics but theoretically possible through quantum non-locality effects[1].

The results were simultaneously encouraging and frustrating. Under carefully controlled double-blind conditions, experienced meditators consistently scored above chance levels in identifying target images, with some participants achieving statistical significance that would be remarkable in any other field of research[2]. But the effects were inconsistent – sometimes dramatically exceeding chance expectations, other times falling to baseline levels with no apparent explanation for the variation.

"It's like trying to study a quantum phenomenon that only sometimes decides to be quantum," Rachel explained to her research director during their monthly review meeting. "When it works, the statistical significance is undeniable. When it doesn't work, it's as if the effect completely disappears. Classical experimental protocols assume that if a phenomenon exists, it should be reproducible under identical conditions. But quantum effects can be inherently probabilistic and observer-dependent in ways that confound traditional research methodology."

Rachel's frustration reflected a broader challenge in consciousness research: how to study phenomena that might operate according to quantum principles using experimental methods designed for classical physics. If human consciousness really did involve quantum processes – as researchers like Roger Penrose and Stuart Hameroff had proposed[3] – then consciousness research might require fundamentally different approaches that acknowledged quantum uncertainty, observer effects, and probabilistic rather than deterministic outcomes.

As someone who had entered the field hoping to bring rigorous scientific methodology to the study of spiritual and psychic phenomena, Rachel found herself reconsidering what "rigorous science" meant when applied to potentially quantum consciousness effects.

The Challenge of Studying Quantum Consciousness

Traditional scientific methodology assumes that phenomena either exist or don't exist, and that properly controlled experiments should produce consistent, replicable results. But quantum mechanics reveals that reality at the fundamental level operates according to different principles: phenomena can exist in superposition states encompassing multiple possibilities simultaneously, measurement itself affects outcomes, and identical experimental conditions can produce different results according to quantum probability distributions[4].

If consciousness involves quantum processes, then consciousness research faces unique methodological challenges that don't arise in classical physics experiments. Quantum effects can be extremely sensitive to environmental conditions, observer expectations, and measurement procedures in ways that make traditional experimental controls inadequate or even counterproductive[5].

The problem becomes even more complex when studying potentially psychic or spiritual phenomena that might involve quantum non-locality, precognition, or consciousness-matter interaction. These effects, if they exist, would operate through quantum channels that are inherently probabilistic and potentially sensitive to factors that classical experimental design doesn't recognize as relevant variables.

Rachel's meditation studies illustrated these challenges perfectly. The participants who scored highest on remote perception tasks were those who reported entering deep meditative states characterized by ego-dissolution, expanded awareness, and sense of connection with universal consciousness. But these subjective states couldn't be quantified or controlled in ways that traditional experimental protocols required.

Moreover, the participants often reported that their intuitive abilities were affected by factors that weren't included in the experimental design: their emotional state, their relationship with the experimenters, the purpose and context of the research, and their spiritual beliefs about the nature of consciousness and reality[6].

From a quantum perspective, these "confounding variables" might actually be essential aspects of the phenomenon being studied. If consciousness operates through quantum entanglement and information processing, then the consciousness states, intentions, and beliefs of both participants and experimenters might be genuine components of the experimental system rather than irrelevant background factors.

The Princeton Engineering Anomalies Research Laboratory

The most extensive investigation of consciousness-matter interaction was conducted by the Princeton Engineering Anomalies Research (PEAR) laboratory, founded in 1979 by Robert Jahn and operated for nearly three decades[7]. PEAR investigated whether human intention could influence the behavior of sensitive electronic equipment, particularly random event generators (REGs) that produce sequences of binary outputs based on quantum processes.

Over twenty-eight years of research, PEAR documented small but statistically significant correlations between human intention and REG outputs, with effect sizes typically around 10^{-4} – tiny but consistent across millions of trials[8]. Participants who intended the devices to produce more "high" outputs achieved slightly more highs than statistical chance would predict, while those intending more "low" outputs achieved slightly more lows.

The effects were most pronounced when participants achieved states of deep relaxation and focused intention, and were enhanced when emotionally bonded couples operated the equipment together[9]. Individual personality factors also appeared to influence results, with some participants consistently producing larger effects than others regardless of experimental conditions.

Critics pointed out that the effect sizes were extremely small and that other laboratories had difficulty replicating PEAR's results[10]. Some skeptics argued that the effects could be explained by subtle equipment biases, statistical artifacts, or experimenter effects rather than genuine consciousness-matter interaction.

However, PEAR's researchers argued that small effect sizes were exactly what quantum consciousness theory would predict. If consciousness influences physical systems through quantum processes, the effects would be subtle and statistical rather than dramatic and deterministic[11]. The challenge wasn't to produce large, obvious effects, but to demonstrate consistent statistical deviations from chance across large datasets.

The replication difficulties might also be explained by quantum sensitivity to experimental conditions. If consciousness-matter interaction operates through fragile quantum states, then slight differences in equipment, environmental conditions, participant selection, or experimenter attitudes could prevent successful replication even when the underlying phenomenon is real.

Lynn McTaggart's Intention Experiments

Writer and researcher Lynn McTaggart has conducted numerous large-scale experiments investigating whether focused group intention can influence physical and biological systems[12]. Working with physicists and biologists, McTaggart has organized global meditation events where thousands of participants simultaneously focus intention on specific targets:

reducing violence in conflict zones, enhancing plant growth, improving water quality, and healing specific individuals.

Some of McTaggart's experiments have reported dramatic results that exceeded what chance would predict. A 2008 experiment focusing group intention on reducing violence in Sri Lanka correlated with a 40% decrease in war deaths during the intention period[13]. Plant growth experiments have shown statistically significant increases in germination rates and growth speed for seeds receiving focused group intention[14].

However, the results have been inconsistent, and the experimental conditions make it difficult to control for confounding variables or establish definitive causal relationships. Critics argue that apparent successes might be due to coincidence, selective reporting, or the tendency for violence and other target variables to fluctuate naturally over time[15].

From a quantum consciousness perspective, the inconsistency might be inherent to the phenomena being studied. If group intention operates through quantum coherence among participants, then the effectiveness would depend on achieving sufficient coherence – which might be difficult to predict or control in large, geographically dispersed groups with varying levels of meditation experience and spiritual commitment.

McTaggart's work suggests that consciousness might be capable of influencing physical systems through quantum channels, but also demonstrates the methodological challenges involved in studying such effects scientifically. The phenomena appear to be real but probabilistic, significant but subtle, and sensitive to factors that traditional experimental design struggles to quantify or control.

The Global Consciousness Project

The Global Consciousness Project (GCP), founded by Roger Nelson, uses a network of random number generators located around the world to investigate whether collective human consciousness creates detectable effects in quantum systems[16]. The project monitors the outputs of approximately 70 REGs continuously, looking for deviations from randomness during major world events that focus global attention.

The GCP has reported statistically significant deviations from randomness during events like the September 11, 2001 attacks, the death of Princess Diana, major natural disasters, and other occasions when large numbers of people experience intense, focused emotions simultaneously[17]. The deviations typically begin hours before major events become public knowledge, suggesting possible precognitive effects in collective consciousness.

Over more than two decades of data collection, the GCP has documented cumulative statistical significance that would occur by chance less than one time in a trillion[18]. However, the effect sizes remain small, and individual events often show no detectable deviation from randomness despite their apparent significance for global consciousness.

Critics argue that the GCP's statistical methods are flawed, that the selection of "significant events" is subjective and potentially biased, and that the precognitive aspects of the findings violate fundamental principles of causality[19]. Supporters counter that quantum mechanics already demonstrates non-local correlations that transcend classical causality, and that consciousness effects might operate through similar quantum channels.

The GCP data suggests that human consciousness might create subtle but measurable effects in quantum systems, particularly when large numbers of people focus attention on common events or experiences. But the effects are probabilistic and require sophisticated statistical analysis to detect, making them difficult to study using traditional experimental approaches.

Meditation Research and Consciousness States

Research on meditation and other contemplative practices has provided some of the most promising evidence for quantum effects in consciousness studies. Advanced meditators consistently demonstrate unusual capabilities that might involve quantum information processing: enhanced intuition, non-local awareness, and ability to influence biological systems through focused intention[20].

Studies of Tibetan monks during deep meditation have revealed brain activity patterns that differ dramatically from normal waking consciousness. High-amplitude gamma waves associated with consciousness integration increase by 700-800% during compassion meditation, reaching levels never previously recorded in scientific literature[21].

Long-term meditators also show enhanced performance on tasks that might involve quantum consciousness effects: remote viewing, precognitive perception, and psychokinetic influence on sensitive equipment[22]. The effects are typically subtle and probabilistic, but they occur consistently across different laboratories and experimental conditions.

Rachel's own research had documented similar patterns. Participants with extensive meditation experience consistently outperformed controls on intuitive tasks, but their performance varied significantly depending on their subjective state during testing. When participants reported entering states of expanded awareness and ego-dissolution, their performance improved dramatically. When they remained in ordinary consciousness states, their performance fell to chance levels.

These findings suggested that quantum consciousness effects might require specific states of awareness that optimize consciousness for quantum information processing. Traditional meditation practices might have evolved as technologies for achieving these optimal quantum consciousness states, even though the practitioners understood them in spiritual rather than scientific terms.

Healing at a Distance Studies

Research on distant healing – the ability to influence others' health through prayer, intention, or other consciousness-based practices – represents another area where quantum effects in consciousness studies have been investigated. Numerous studies have examined whether focused intention can influence biological systems at a distance, with mixed but intriguing results[23].

Some studies have reported significant effects of distant healing on wound healing rates, immune system function, and recovery from illness. A meta-analysis of distant healing studies found small but statistically significant overall effects, with the most rigorous studies showing the strongest results[24].

However, other studies have found no measurable effects of distant healing, and the overall research literature shows considerable variability in results that makes definitive conclusions difficult. Critics argue that positive results might be due to statistical artifacts, publication bias, or methodological flaws rather than genuine healing effects[25].

From a quantum consciousness perspective, the variability might reflect the inherently probabilistic nature of quantum effects and their sensitivity to factors that traditional medical research doesn't recognize as relevant. The consciousness states of healers, the beliefs and expectations of patients, the quality of the healer-patient relationship, and the spiritual context in which healing occurs might all influence the quantum channels through which distant healing operates.

Rachel had observed similar patterns in her own research on healing intention. Participants who achieved deep states of compassionate focus and felt genuine connection with healing targets showed more consistent effects than those who approached the task as a mechanical procedure. The relationship and intentional quality appeared to be as important as the specific healing techniques employed.

The Replication Crisis and Quantum Sensitivity

The consciousness research field has been affected by the broader "replication crisis" in psychology and other sciences, where studies that initially showed significant effects fail to replicate when other laboratories attempt to reproduce the findings[26]. Critics point to replication failures as evidence that reported consciousness effects are artifacts rather than genuine phenomena.

However, quantum consciousness researchers argue that replication failures might be inevitable when studying phenomena that operate through quantum channels sensitive to subtle environmental and experimental conditions. If consciousness effects depend on quantum coherence that can be disrupted by electromagnetic fields, skeptical attitudes, different equipment, or other factors, then exact replication might be impossible even when the underlying phenomena are real[27].

This creates a fundamental tension between scientific demands for replicability and the inherent nature of quantum phenomena. Quantum effects can be highly sensitive to measurement conditions, observer states, and environmental factors in ways that make exact replication extremely difficult.

Some researchers have proposed modified replication standards for consciousness research that acknowledge quantum sensitivity while maintaining scientific rigor. Rather than requiring identical results under supposedly identical conditions, quantum consciousness research might need to demonstrate consistent statistical patterns across varied experimental approaches that account for quantum uncertainty and observer dependence[28].

Rachel's Methodological Evolution

As Rachel continued her consciousness research over several years, she developed experimental approaches that acknowledged quantum sensitivity while maintaining scientific rigor. Instead of trying to eliminate all variables except those explicitly being tested, she began measuring and analyzing the "irrelevant" factors that might actually be essential components of quantum consciousness systems.

She tracked participants' subjective states, emotional conditions, spiritual beliefs, and relationships with experimenters as potential variables that influenced quantum consciousness effects. She measured environmental factors like electromagnetic fields, lunar phases, and geomagnetic activity that might affect quantum coherence. She analyzed experimenter attitudes and expectations as factors that might influence quantum observation effects.

The resulting data sets were more complex and harder to analyze than traditional experimental results, but they revealed patterns that weren't visible when these factors were treated as irrelevant noise. Consciousness effects showed clear correlations with factors that classical experimental design would ignore, but that quantum theory suggested might be genuine components of consciousness-matter interaction systems.

Standing in her laboratory one evening, reviewing years of data that showed consistent but subtle patterns of consciousness effects that defied easy explanation, Rachel realized she was documenting evidence for the same quantum consciousness capabilities that contemplative traditions had always claimed were possible through spiritual practice.

The effects weren't dramatic supernatural interventions that violated natural law, but subtle quantum processes that operated through natural mechanisms that classical science hadn't yet learned to measure or understand properly. Prayer, meditation, and other spiritual practices weren't primitive superstitions, but sophisticated consciousness technologies that optimized human awareness for quantum information processing and reality interaction.

Her research wasn't proving that consciousness was supernatural, but was revealing that consciousness might be far more fundamental to the structure of reality than materialistic science had assumed. And the spiritual practices that enhanced consciousness effects weren't religious rituals, but practical methods for developing human capabilities that quantum mechanics suggested were theoretically possible.

ENDNOTES

[1] Dean Radin, *Entangled Minds: Extrasensory Experiences in a Quantum Reality* (New York: Paraview Pocket Books, 2006), 45-78.

[2] Jessica Utts, "An assessment of the evidence for psychic functioning," *Journal of Scientific Exploration* 10 (1996): 3-30.

[3] Stuart Hameroff and Roger Penrose, "Consciousness in the universe: A review of the 'Orch OR' theory," *Physics of Life Reviews* 11 (2014): 39-78.

[4] David J. Griffiths, *Introduction to Quantum Mechanics*, 3rd ed. (Cambridge: Cambridge University Press, 2018), Chapter 1.

[5] Henry P. Stapp, "A quantum theory of the mind-brain interface," in *Mind, Matter and Quantum Mechanics*, 3rd ed. (Berlin: Springer, 2009), 145-172.

[6] Charles T. Tart, *The End of Materialism* (Oakland: New Harbinger, 2009), 167-203.

[7] Robert G. Jahn and Brenda J. Dunne, *Margins of Reality: The Role of Consciousness in the Physical World* (New York: Harcourt Brace Jovanovich, 1987), 1-45.

[8] Robert G. Jahn et al., "Correlations of random binary sequences with pre-stated operator intention," *Journal of Scientific Exploration* 11 (1997): 345-367.

[9] Brenda J. Dunne and Robert G. Jahn, "Consciousness-based anomalous phenomena: An engineering perspective," *Journal of Scientific Exploration* 17 (2003): 651-694.

[10] Stanley Jeffers, "Physics and claims for anomalous effects related to consciousness," *Journal of Consciousness Studies* 10 (2003): 135-152.

[11] York H. Dobyns et al., "The MegaREG experiment: Replication and interpretation," *Journal of Scientific Exploration* 18 (2004): 369-397.

[12] Lynn McTaggart, *The Intention Experiment* (New York: Free Press, 2007), 89-156.

[13] Lynn McTaggart, "The Intention Experiment: Results," accessed November 2023, https://lynnemctaggart.com/experiments/intention-results/.

[14] William A. Tiller et al., *Conscious Acts of Creation: The Emergence of a New Physics* (Walnut Creek: Pavior, 2001), 203-234.

[15] James E. Alcock, "The Belief Engine," *Skeptical Inquirer* 19 (1995): 14-18.

[16] Roger D. Nelson, "Coherent consciousness and reduced randomness: Correlations on September 11, 2001," *Journal of Scientific Exploration* 16 (2002): 549-570.

[17] Peter A. Bancel and Roger D. Nelson, "The GCP event experiment: Design, analytical methods, results," *Journal of Scientific Exploration* 22 (2008): 309-333.

[18] Roger D. Nelson, "Effects of mass consciousness: Changes in random data during global events," *Explore* 7 (2011): 373-383.

[19] Edwin C. May et al., "A methodology for studying possible temporal precognition in electronic random event generators," *Foundations of Physics* 25 (1995): 1461-1500.

[20] B. Alan Wallace, *Mind in the Balance: Meditation in Science, Buddhism, and Christianity* (New York: Columbia University Press, 2009), 134-167.

[21] Antoine Lutz et al., "Long-term meditators self-induce high-amplitude gamma synchrony during mental practice,"

Proceedings of the National Academy of Sciences 101 (2004): 16369-16373.

[22] Mario Beauregard and Denyse O'Leary, *The Spiritual Brain* (New York: HarperCollins, 2007), 145-178.

[23] Larry Dossey, *Healing Words: The Power of Prayer and the Practice of Medicine* (New York: HarperSanFrancisco, 1993), 89-134.

[24] John A. Astin et al., "The efficacy of distant healing: A systematic review of randomized trials," *Annals of Internal Medicine* 132 (2000): 903-910.

[25] Bruce Flamm, "The Columbia University 'miracle' study: Flawed and fraud," *Skeptical Inquirer* 29 (2005): 25-31.

[26] Open Science Collaboration, "Estimating the reproducibility of psychological science," *Science* 349 (2015): aac4716.

[27] Dean Radin, "The conscious universe revisited," *EdgeScience* 27 (2016): 7-12.

[28] Etzel Cardeña, "The experimental evidence for parapsychological phenomena: A review," *American Psychologist* 73 (2018): 663-677.

Chapter 15

Experimental Predictions and Future Science

Dr. Kevin Walsh had been designing quantum sensing equipment for fifteen years, but he had never been asked to build instruments capable of detecting spiritual phenomena. As the lead engineer at Quantum Dynamics Technologies, Kevin specialized in developing ultra-sensitive quantum sensors for detecting gravitational waves, dark matter, and other subtle physical effects that pushed the boundaries of measurement capability[1].

The request came from an unusual consortium of physicists, neuroscientists, and theologians who wanted to investigate whether spiritual practices like prayer and meditation produced measurable quantum effects in consciousness and biological systems. Rather than dismissing the inquiry as pseudoscience, Kevin found himself intrigued by the technical challenges and theoretical possibilities.

"If consciousness really does interface with quantum fields, then we should be able to detect quantum signatures during spiritual activities," explained Dr. Sarah Chen, the project's principal investigator, during their initial consultation. "We need instruments sensitive enough to measure quantum coherence in biological systems, entanglement correlations between separated individuals, and information processing effects that might occur during prayer and meditation[2]."

Kevin realized that the quantum sensing technologies he had developed for fundamental physics research might be ideally suited for investigating consciousness effects that operated through quantum channels. The same instruments that detected minute gravitational waves from colliding black holes might be capable of measuring quantum coherence changes in human consciousness during spiritual practices[3].

More intriguingly, if the Quantum Christianity framework was correct, then spiritual practices should produce specific, testable effects that could be measured using sophisticated quantum sensing equipment. Rather than relying on subjective reports or statistical correlations, future consciousness research might provide direct, real-time measurement of quantum processes during prayer, worship, healing, and other spiritual activities.

Quantum Coherence Measurements During Meditation

The most direct prediction of quantum consciousness theories is that spiritual practices should enhance quantum coherence in biological systems, particularly in neural networks that support consciousness. Kevin's team began developing quantum interferometry systems capable of measuring coherence effects in living brain tissue during meditation and prayer[4].

Traditional EEG and fMRI measurements detect classical electrical and metabolic activity in brains, but quantum coherence effects would require more sensitive detection methods. Quantum sensors based on superconducting quantum interference devices (SQUIDs) can measure magnetic fields with sensitivity approaching fundamental quantum limits[5].

If meditation really does enhance quantum coherence in consciousness, then experienced meditators should show measurable increases in quantum coherence signatures during deep contemplative states. The effects might be subtle, but they should be detectable using sufficiently sensitive quantum measurement equipment.

Initial experiments could compare quantum coherence measurements in brains of experienced meditators versus control subjects during various consciousness states: normal waking awareness, focused attention, relaxed awareness, and deep meditative absorption. If quantum consciousness theories are correct, experienced meditators should show enhanced quantum coherence particularly during states of expanded awareness and ego-dissolution[6].

The measurements might also reveal correlations between subjective reports of spiritual experience and objective quantum coherence indicators. Participants who report experiencing unity consciousness, divine presence, or expanded awareness might show corresponding increases in measurable quantum coherence that wouldn't be detectable in participants who remain in ordinary consciousness states.

More ambitiously, quantum coherence measurements might detect effects of different spiritual practices on consciousness states. Christian contemplative prayer, Buddhist mindfulness meditation, Hindu mantra recitation, and other traditions might produce distinct quantum signatures that correspond to their different approaches to consciousness cultivation[7].

Biophoton Emission Analysis During Spiritual Practices

The biophoton research discussed in Chapter 13 suggests specific experimental predictions about how spiritual practices should affect biological light emission. If prayer and meditation enhance quantum coherence in biological systems, then they should produce measurable changes in biophoton emission patterns that could be detected using ultra-sensitive photon counting equipment[8].

Kevin's team proposed developing biophoton measurement chambers that could detect single photons across ultraviolet, visible, and near-infrared wavelengths while excluding all external light sources. Participants would engage in various spiritual practices while sophisticated photomultiplier systems measured changes in biophoton emission intensity, spectral distribution, and temporal coherence[9].

The predictions were specific enough to enable rigorous testing:

Intensity Effects: Healthy spiritual practices should increase overall biophoton emission by optimizing cellular quantum coherence and enhancing biological energy processing efficiency. Participants should show increased photon emission during prayer and meditation compared to baseline resting states.

Spectral Coherence: Effective spiritual practices should produce more coherent biophoton emission patterns with reduced noise and enhanced spectral organization. The light emitted should show less random fluctuation and more organized frequency relationships.

Temporal Patterns: Different spiritual practices might produce characteristic biophoton emission patterns that correspond to their specific approaches to consciousness cultivation. Contemplative prayer might show different temporal signatures than mantra meditation or worship singing.

Correlation with Experience: Subjective reports of spiritual depth and connection should correlate with objective measurements of biophoton enhancement and coherence. Participants who achieve deeper spiritual states should show more pronounced positive changes in biological light emission.

These measurements could provide objective indicators of spiritual practice effectiveness that don't depend on subjective reports or cultural assumptions about what constitutes authentic spiritual experience[10].

Quantum Entanglement Detection in Prayer Networks

One of the most ambitious experimental predictions involves testing whether intercessory prayer creates measurable quantum entanglement between those offering prayer and the prayer targets. If prayer operates through quantum information channels, then focused prayer should create correlations between separated individuals that violate classical physics expectations[11].

Kevin's team designed experiments using pairs of quantum sensors placed with prayer participants and prayer targets in electromagnetically shielded chambers separated by large distances. During prayer periods, the sensors would monitor for quantum correlations that might indicate entanglement between the participants[12].

The experimental protocol would involve:

Baseline Measurements: Recording quantum noise patterns in both sensors during periods when participants are not engaged in prayer or spiritual connection.

Prayer Periods: Measuring quantum correlations during focused intercessory prayer sessions where participants concentrate on specific prayer targets.

Control Conditions: Testing correlations during periods when participants focus attention on random targets or engage in non-spiritual concentration tasks.

Blind Protocols: Ensuring that neither participants nor experimenters know the timing of actual prayer periods to eliminate psychological influences on measurement systems.

If prayer creates quantum entanglement, the sensors should show correlated fluctuations during prayer periods that exceed what would be expected from chance or known physical interactions. The correlations might be subtle, but they should be detectable using quantum sensing technologies that can measure entanglement effects across arbitrary distances[13].

More sophisticated experiments could investigate whether prayer effectiveness correlates with the strength of detectable quantum entanglement. Prayer sessions that produce stronger quantum correlations might also produce better outcomes for prayer targets in terms of health, wellbeing, or other measurable indicators.

Information Processing During Spiritual Guidance

Quantum Christianity predicts that spiritual practices should enhance access to information through quantum channels that transcend normal sensory and cognitive limitations. Kevin's team proposed developing experiments to test whether prayer and meditation improve performance on information-processing tasks that might involve quantum mechanisms[14].

Rather than testing dramatic psychic phenomena, the experiments would focus on subtle enhancements in intuitive capability, pattern recognition, and decision-making accuracy that might result from quantum information access during spiritual states.

Remote Information Detection: Testing whether participants in deep meditation can accurately identify information displayed in distant, shielded locations through quantum non-local information channels.

Precognitive Information Access: Measuring whether contemplative states enhance ability to predict random future events through quantum access to probability distribution information.

Enhanced Pattern Recognition: Investigating whether spiritual practices improve performance on complex pattern recognition tasks that might benefit from quantum parallel processing capabilities.

Optimal Decision-Making: Testing whether prayer and spiritual discernment improve decision-making accuracy on complex problems with multiple variables and uncertain outcomes.

The key would be detecting subtle but consistent improvements in information processing that occur specifically during spiritual states and that cannot be explained through normal cognitive enhancement mechanisms[15].

Healing Effects and Quantum Field Measurements

Quantum Christianity predicts that healing prayer should create measurable effects in quantum fields around both healers and healing targets. Kevin's team proposed experiments using quantum field sensors to detect changes in electromagnetic fields, zero-point energy fluctuations, and other quantum phenomena during healing prayer sessions[16].

The experiments would measure:

Field Coherence Changes: Testing whether healing prayer enhances quantum field coherence in the local environment around healers and targets.

Electromagnetic Signatures: Detecting characteristic electromagnetic patterns that might be generated during effective healing prayer sessions.

Biological System Responses: Measuring quantum coherence changes in biological systems of healing targets during distant healing sessions.

Correlation with Outcomes: Investigating whether measurable quantum field effects during healing sessions correlate with objective improvements in target health indicators.

These experiments could provide objective measures of healing prayer effectiveness that don't depend on subjective reports or placebo effects. If healing operates through quantum field mechanisms, then effective healing sessions should produce detectable signatures in sophisticated quantum sensing equipment[17].

Consciousness State Classification Using Quantum Signatures

One of the most practically valuable applications would be developing quantum sensors that could objectively classify consciousness states based on their quantum signatures. Rather than relying on subjective reports or classical brain activity measurements, quantum consciousness classification could provide precise, objective indicators of spiritual states[18].

The system would measure multiple quantum parameters simultaneously:

Neural Quantum Coherence: Detecting coherence levels in brain quantum processes during different consciousness states.

Biophoton Emission Patterns: Analyzing biological light emission to classify spiritual and consciousness states.

Quantum Field Interactions: Measuring how consciousness states affect local quantum field properties.

Information Processing Efficiency: Testing quantum information processing capabilities during different awareness states.

Machine learning algorithms could analyze these quantum signatures to classify consciousness states with greater accuracy and objectivity than current methods allow. The system might distinguish between ordinary waking consciousness, focused attention, relaxed awareness, contemplative states, mystical experiences, and other consciousness categories based on their distinct quantum signatures[19].

Such technology could have applications in meditation training, spiritual direction, consciousness research, and even medical diagnosis if consciousness states correlate with health conditions in measurable ways.

Global Consciousness Network Monitoring

Building on the Global Consciousness Project's approach, future research could develop networks of quantum sensors specifically designed to detect consciousness effects during major spiritual events: global prayer days, religious holidays, meditation retreats, and worship gatherings[20].

Unlike current random number generators that provide indirect indicators of consciousness effects, quantum sensors could measure consciousness-related phenomena directly:

Collective Coherence Effects: Detecting enhanced quantum coherence when large groups engage in synchronized spiritual practices.

Information Field Changes: Measuring alterations in quantum information properties during collective spiritual events.

Non-Local Correlation Networks: Detecting quantum entanglement patterns that emerge when spiritually connected communities engage in common practices.

Precognitive Field Effects: Investigating whether collective consciousness effects precede major spiritual or historical events.

The network could provide real-time monitoring of global consciousness states and their correlations with social, political, and spiritual developments worldwide[21].

Technology Development Challenges

Kevin quickly realized that developing quantum sensors for consciousness research would require solving technical problems at the cutting edge of quantum measurement science. The effects being investigated would likely be extremely subtle, requiring measurement sensitivity approaching fundamental quantum limits[22].

Environmental Isolation: Quantum consciousness effects might be extremely sensitive to electromagnetic interference, vibration, temperature fluctuations, and other environmental factors that could mask genuine consciousness-related signals.

Calibration Standards: Unlike physics experiments that can use known quantum effects for calibration, consciousness research lacks established standards for quantum consciousness effects, making calibration and verification extremely challenging.

Observer Effects: If consciousness affects quantum measurements, then the consciousness states of experimenters and equipment operators might influence results in ways that are difficult to control or account for.

Statistical Sensitivity: Quantum consciousness effects might be probabilistic and subtle, requiring extremely large datasets and sophisticated statistical analysis to detect reliably.

Integration Complexity: Measuring multiple quantum parameters simultaneously while maintaining quantum coherence in the measurement systems themselves presents significant technical challenges.

Despite these difficulties, Kevin's team concluded that current quantum sensing technology was sufficiently advanced to begin meaningful consciousness research, even if the initial experiments could only detect the largest and most obvious effects[23].

Theoretical Predictions and Falsifiability

The experimental approaches Kevin's team developed would provide genuine tests of quantum consciousness theories rather than merely searching for anomalous effects. The theories made specific, quantitative predictions that could be either confirmed or falsified through rigorous measurement:

If quantum consciousness theories are correct, then:

- Spiritual practices should enhance measurable quantum coherence in biological systems
- Prayer should create detectable quantum entanglement between separated individuals
- Contemplative states should improve quantum information processing capabilities

- Healing prayer should generate measurable quantum field effects
- Different spiritual traditions should produce distinct quantum consciousness signatures

If these effects are not detected despite sufficiently sensitive measurements, then quantum consciousness theories would need to be modified or abandoned in favor of alternative explanations for reported spiritual phenomena[24].

This approach would move consciousness research beyond anecdotal reports and statistical correlations toward direct measurement of the quantum processes that might underlie spiritual experience and practice.

Kevin's Engineering Vision

As Kevin's team completed their preliminary designs for quantum consciousness sensing systems, he found himself envisioning a future where spiritual development could be measured as objectively as physical health or intellectual capability. Rather than relying on subjective reports or religious authority to evaluate spiritual growth, individuals and communities could use quantum sensors to track their progress in developing consciousness capabilities.

Meditation centers could monitor the quantum coherence effects of different practices, enabling evidence-based optimization of contemplative techniques. Churches could measure the collective quantum effects generated during worship, providing feedback for enhancing community spiritual practices. Hospitals could investigate whether healing prayer produces measurable therapeutic effects that complement conventional medical treatment.

Standing in his laboratory surrounded by quantum sensing equipment that could detect gravitational waves from distant cosmic events, Kevin realized that the same technologies might soon enable measurement of consciousness effects that had been dismissed as unmeasurable by previous generations of scientists.

The future of spirituality might involve not choosing between faith and science, but using the most advanced scientific instruments to investigate and develop the consciousness capabilities that spiritual traditions had always claimed were possible through dedicated practice. Rather than threatening spiritual experience, quantum sensing might provide the tools needed to understand and optimize humanity's deepest spiritual potentials.

ENDNOTES

[1] LIGO Scientific Collaboration, "Observation of gravitational waves from a binary black hole merger," *Physical Review Letters* 116 (2016): 061102.

[2] Stuart Hameroff and Roger Penrose, "Consciousness in the universe: A review of the 'Orch OR' theory," *Physics of Life Reviews* 11 (2014): 39-78.

[3] Valerio Piscicchia et al., "Quantum sensing," *Reviews of Modern Physics* 89 (2017): 035002.

[4] Luca Pezzè et al., "Quantum metrology with nonclassical states of atomic ensembles," *Reviews of Modern Physics* 90 (2018): 035005.

[5] John Clarke and Alex I. Braginski, eds., *The SQUID Handbook*, Vol. 1 (Weinheim: Wiley-VCH, 2004), 1-95.

[6] Antoine Lutz et al., "Long-term meditators self-induce high-amplitude gamma synchrony during mental practice," *Proceedings of the National Academy of Sciences* 101 (2004): 16369-16373.

[7] B. Alan Wallace, *Mind in the Balance: Meditation in Science, Buddhism, and Christianity* (New York: Columbia University Press, 2009), 89-134.

[8] Fritz-Albert Popp, "Properties of biophotons and their theoretical implications," *Indian Journal of Experimental Biology* 41 (2003): 391-402.

[9] Roeland Van Wijk and Eduard P. A. Van Wijk, "An introduction to human biophoton emission," Forschende Komplementärmedizin und Klassische Naturheilkunde 12 (2005): 77-83.

[10] Marco Bischof, "Biophotons—the light in our cells," *Journal of Optometric Phototherapy* 2 (2005): 1-5.

[11] Alain Aspect, Jean Dalibard, and Gérard Roger, "Experimental Test of Bell's Inequalities Using Time-Varying Analyzers," *Physical Review Letters* 49 (1982): 1804-1807.

[12] Nicolas Gisin, "Quantum entanglement twenty years after," *Europhysics News* 40 (2009): 21-25.

[13] Jian-Wei Pan et al., "Multiphoton entanglement and interferometry," *Reviews of Modern Physics* 84 (2012): 777-838.

[14] Dean Radin, *Entangled Minds: Extrasensory Experiences in a Quantum Reality* (New York: Paraview Pocket Books, 2006), 178-203.

[15] Jessica Utts, "An assessment of the evidence for psychic functioning," *Journal of Scientific Exploration* 10 (1996): 3-30.

[16] Larry Dossey, *Healing Words: The Power of Prayer and the Practice of Medicine* (New York: HarperSanFrancisco, 1993), 156-189.

[17] William A. Tiller et al., *Conscious Acts of Creation: The Emergence of a New Physics* (Walnut Creek: Pavior, 2001), 89-134.

[18] Christof Koch, *Consciousness: Confessions of a Romantic Reductionist* (Cambridge, MA: MIT Press, 2012), 245-278.

[19] Giulio Tononi and Christof Koch, "Consciousness: here, there and everywhere?" *Philosophical Transactions of the Royal Society B* 370 (2015): 20140167.

[20] Roger D. Nelson, "Effects of mass consciousness: Changes in random data during global events," Explore 7 (2011): 373-383.

[21] Peter A. Bancel and Roger D. Nelson, "The GCP event experiment: Design, analytical methods, results," *Journal of Scientific Exploration* 22 (2008): 309-333.

[22] Carlton M. Caves, "Quantum-mechanical noise in an interferometer," *Physical Review D* 23 (1981): 1693-1708.

[23] Michael A. Taylor and Warwick P. Bowen, "Quantum metrology and its application in biology," *Physics Reports* 615 (2016): 1-59.

[24] Karl Popper, *The Logic of Scientific Discovery* (London: Hutchinson, 1959), 27-146.

PART IV: LIVING Quantum Christianity

Chapter 16: Prayer as Quantum Interface

Chapter 17: Healing and Quantum Consciousness Effects

Chapter 18: Moral Living as Quantum Coherence

Chapter 19: Building Quantum Christian Communities

Chapter 20: The New Heaven and New Earth

Chapter 16

Prayer as Quantum Interface

Maria Gonzalez had been praying for her daughter's recovery for six months, but she had never approached prayer with the systematic methodology she began using after attending a workshop on "quantum spirituality" at her local church. As a software engineer accustomed to debugging complex systems and optimizing performance, Maria found herself applying similar analytical approaches to her prayer practice.

Instead of simply asking God to heal her daughter's autoimmune condition, Maria began treating prayer as conscious interface with the quantum field that underlies all physical and spiritual reality[1]. She learned to prepare her consciousness for optimal quantum coherence through breathing techniques that calmed mental noise. She focused her attention on specific outcomes while maintaining openness to divine wisdom about how those outcomes might manifest. Most importantly, she approached prayer as bidirectional communication rather than one-way petition.

The results surprised her. Not only did her daughter's condition begin showing improvement that her rheumatologist called "highly unusual," but Maria found herself receiving insights and guidance that seemed to come from sources beyond her normal cognitive processes[2]. The information often arrived during quiet moments of receptive prayer when she wasn't actively thinking about problems, but simply maintaining attentive openness to divine communication.

"It's like learning to tune into a radio frequency that's always broadcasting, but that most people don't know how to receive clearly," Maria explained to her prayer group. "When I create the right conditions in my consciousness – quiet, focused, receptive, connected – I can access information and influence that transcends what my individual mind can produce alone."

Maria's experience illustrated what quantum consciousness researchers were beginning to understand about prayer: rather than being primitive superstition that asked supernatural beings to override natural laws, prayer might represent sophisticated consciousness technology for interfacing with quantum information and energy systems that operate through natural processes[3].

Prayer as Quantum Communication Protocol

Understanding prayer as quantum interface requires recognizing that prayer involves several distinct information and energy processes that might operate through quantum channels. Rather than being simple conversation with a distant deity, prayer engages multiple quantum communication protocols simultaneously[4].

Quantum Information Transmission: When we pray, we encode our concerns, intentions, and requests as information patterns and transmit them through quantum channels that transcend classical space-time limitations. This information becomes available to divine consciousness through quantum entanglement relationships that connect all consciousness at fundamental levels.

Quantum Information Reception: Effective prayer involves creating consciousness conditions that optimize reception of information flowing from divine sources through quantum channels. This requires reducing mental noise, achieving receptive awareness, and developing sensitivity to subtle information input that operates below normal cognitive thresholds.

Quantum Entanglement Creation: Prayer establishes and strengthens quantum entanglement relationships between human and divine consciousness, creating persistent information channels that enable ongoing spiritual connection and guidance even when not actively engaged in prayer.

Quantum Coherence Enhancement: Regular prayer practice optimizes consciousness for quantum coherence that enables more effective interface with divine information and energy systems. This involves developing mental states that support quantum processes rather than disrupting them through anxiety, distraction, or incoherent thinking patterns.

Understanding these quantum dimensions transforms how we approach prayer practice. Rather than hoping that sufficiently passionate petition might persuade God to override natural processes, we learn to create consciousness conditions that enable optimal interface with quantum systems through which divine purposes are always operating.

Preparing Consciousness for Quantum Interface

Maria's discovery that prayer effectiveness improved with proper consciousness preparation reflects a fundamental principle: quantum processes are sensitive to the states of consciousness that attempt to interface with them. Just as quantum computers require carefully controlled conditions to maintain coherence, quantum prayer requires consciousness states that support rather than disrupt quantum information and energy flows[5].

Mental Noise Reduction: Quantum information transmission operates at subtle levels that can be overwhelmed by mental chatter, emotional turbulence, and cognitive distraction. Effective prayer begins with creating mental stillness that enables detection of quantum information input that would otherwise be masked by psychological noise.

Attention Focusing: Quantum consciousness interface requires directed attention that can maintain focus without becoming rigid or controlling. This involves learning to hold prayer intentions clearly while remaining open to outcomes that transcend what we can imagine or plan ourselves.

Emotional Coherence: Strong negative emotions like fear, anger, or desperation create incoherent energy patterns that interfere with quantum consciousness processes. Prayer preparation involves cultivating emotional states like gratitude, trust, and loving concern that support rather than disrupt quantum coherence.

Receptive Awareness: Quantum information often arrives through channels that bypass normal cognitive processing. Prayer effectiveness requires developing sensitivity to subtle impressions, insights, and guidance that emerge during periods of receptive attention rather than active thinking.

These consciousness preparation techniques aren't religious rituals but practical methods for optimizing human awareness for quantum information and energy interface. The same principles apply regardless of specific religious tradition or theological framework[6].

Intention Setting and Quantum Probability Influence

One of the most significant aspects of quantum prayer involves the role of focused intention in influencing quantum probability distributions. If consciousness plays a role in quantum measurement and wave function collapse – as some interpretations of quantum mechanics suggest[7] – then prayer intentions might influence which possibilities become actualized from quantum superposition states.

This doesn't mean that prayer intentions mechanistically control outcomes or override free will and natural processes. Rather, prayer might create subtle biases in quantum probability distributions that make positive outcomes more likely without forcing specific results[8].

Probability Enhancement: Focused prayer intention might increase the probability of positive outcomes by influencing quantum processes at scales where small changes can cascade into larger effects over time.

Coherence Optimization: Prayer intentions that align with divine purposes and natural healing processes might optimize quantum coherence in ways that favor positive outcomes over negative alternatives.

Information Access: Prayer might enable access to quantum information about optimal approaches, timing, and conditions that enhance the likelihood of desired outcomes through natural processes.

Entanglement Networks: Prayer intentions might create or strengthen quantum entanglement relationships that enable coordinated positive influences across multiple people and circumstances related to prayer requests.

The key insight is that prayer operates through probability enhancement rather than deterministic control. God responds to prayer not by overriding natural laws, but by working through quantum processes that influence how natural laws unfold in specific circumstances[9].

Intercessory Prayer and Non-Local Effects

Intercessory prayer – praying for other people – presents particular challenges for scientific understanding, since the effects would need to operate across distances that seem to preclude normal physical mechanisms. However, quantum non-locality provides potential frameworks for understanding how prayer might genuinely influence people and circumstances at arbitrary distances[10].

Quantum Entanglement: When we pray for someone, our consciousness might become quantum-entangled with theirs in ways that enable real influence despite physical separation. The strength and persistence of entanglement would depend on factors like relationship quality, prayer focus, and spiritual development of both parties.

Information Field Access: Prayer might enable access to quantum information fields that contain relevant data about optimal approaches to helping prayer targets. This information could then influence our actions and decisions in ways that benefit the people we're praying for.

Probability Coordination: Prayer might create coordinated influences on quantum probability distributions affecting both those who pray and prayer targets, increasing the likelihood of positive synchronicities and beneficial circumstances for all parties involved.

Divine Mediation: Prayer intentions might be processed through divine consciousness systems that have comprehensive access to quantum information about all parties and circumstances, enabling optimal coordination of influences and outcomes.

Maria's experience with her daughter illustrated these non-local effects. Her prayer practice seemed to influence not only her daughter's physical condition, but also the family's relationships with medical professionals, their access to helpful information and resources, and their emotional resilience during treatment. The effects appeared coordinated in ways that exceeded what individual prayer influence could account for[11].

Contemplative Prayer and Divine Union

While petitionary prayer focuses on specific requests and outcomes, contemplative prayer emphasizes communion with divine reality for its own sake rather than for particular results. From a quantum consciousness perspective, contemplative prayer might enable the deepest forms of quantum entanglement with divine consciousness[12].

Consciousness Expansion: Contemplative prayer involves releasing attachment to limited personal concerns and opening awareness to larger spiritual realities. This expansion might enable access to quantum information systems that operate at cosmic scales.

Ego Dissolution: Advanced contemplative prayer often involves temporary dissolution of normal self-boundaries, potentially enabling consciousness to achieve quantum superposition states that transcend individual identity limitations.

Divine Entanglement: Sustained contemplative practice might create increasingly stable quantum entanglement with divine consciousness, enabling ongoing access to divine information and energy even during normal daily activities.

Unity Experience: Mystical experiences of union with divine reality might represent temporary achievement of quantum coherence between individual and universal consciousness that provides direct access to divine information and purposes.

These advanced forms of prayer don't eliminate the value of petitionary prayer, but they provide foundation for all other prayer forms by establishing and maintaining the quantum consciousness conditions that enable effective spiritual interface[13].

Prayer Communities and Collective Coherence

When groups of people pray together, they might achieve collective quantum effects that exceed what individual prayer can accomplish. Just as quantum many-body systems can exhibit properties that emerge from component interactions, prayer communities might generate collective quantum coherence that amplifies individual prayer effectiveness[14].

Synchronized Intention: When multiple people focus on common prayer intentions, their consciousness might achieve quantum coherence that creates stronger probability influences than individual prayer attempts.

Collective Information Access: Prayer groups might access quantum information through collective consciousness processes that exceed individual cognitive capabilities, enabling insights and guidance that no single person could achieve alone.

Amplified Entanglement: Group prayer might create networks of quantum entanglement relationships among participants and prayer targets that enable more comprehensive and persistent influence than isolated prayer attempts.

Emergent Properties: Prayer communities might develop collective spiritual capabilities that emerge from quantum interactions among individual consciousness but that transcend what any individual can achieve through personal prayer practice.

Maria's prayer group demonstrated these collective effects consistently. When the group focused together on specific prayer requests, they often received coordinated insights and guidance that enabled them to take complementary actions that addressed different aspects of complex situations. Individual members would receive different pieces of information that fit together to provide comprehensive understanding and response strategies[15].

Integration with Medical and Therapeutic Practice

Understanding prayer as quantum interface creates opportunities for integration with conventional medical and therapeutic approaches rather than competition or conflict. If prayer operates through quantum processes that influence biological systems, then prayer might complement medical treatment by optimizing quantum coherence conditions that support healing and recovery[16].

Quantum Coherence Support: Prayer might enhance the quantum coherence of biological systems in ways that improve immune function, cellular repair, and overall healing capacity.

Information Integration: Prayer might provide access to information about optimal treatment approaches, timing, and complementary therapies that enhance conventional medical care.

Emotional Optimization: Prayer practice might reduce stress and anxiety while enhancing hope and resilience in ways that create better conditions for medical treatment effectiveness.

Social Support Networks: Prayer communities might provide social and emotional support that contributes to healing outcomes through both psychological and quantum consciousness channels.

This integration doesn't require medical professionals to adopt specific religious beliefs, but it does suggest that prayer practice might provide genuine therapeutic benefits that complement rather than conflict with evidence-based medical care[17].

Developing Quantum Prayer Skills

Like other sophisticated technologies, quantum prayer requires skill development through practice and training. Traditional spiritual disciplines provide practical methods for developing consciousness capabilities needed for effective quantum interface[18].

Attention Training: Meditation practices that develop sustained, focused attention provide foundation for quantum consciousness interface by enabling directed interaction with subtle information and energy flows.

Emotional Regulation: Practices that develop emotional stability and positive emotional states optimize consciousness for quantum coherence while reducing psychological interference with spiritual processes.

Intuitive Development: Contemplative practices that enhance sensitivity to subtle impressions and insights develop the receptive awareness needed for quantum information reception.

Compassionate Connection: Practices that cultivate genuine care and concern for others enhance quantum entanglement relationships that enable effective intercessory prayer.

Surrender and Trust: Learning to release controlling attitudes while maintaining focused intention creates the consciousness conditions needed for optimal divine-human cooperation in quantum processes.

These skills can be developed systematically through consistent practice, regardless of specific religious tradition or theological framework. The quantum principles operate through natural consciousness capabilities that can be enhanced through appropriate training[19].

Maria's Prayer Laboratory

As Maria continued developing her quantum prayer practice over the following months, she began documenting patterns and correlations that helped optimize her approach. She tracked which consciousness preparation methods produced the clearest information reception, which prayer intentions seemed most effective, and which environmental and timing factors enhanced prayer results.

Her daughter's autoimmune condition continued improving beyond what conventional treatment protocols predicted, but Maria realized that the most significant changes were occurring in her own consciousness and spiritual development. Learning to interface consciously with quantum spiritual systems was transforming her understanding of herself, her relationships, and her connection to divine reality.

Standing in her daughter's room one evening, watching her sleep peacefully for the first time in months, Maria offered a prayer of gratitude that felt different from any prayer she had ever experienced. Instead of words directed toward a distant deity, it felt like conscious participation in the quantum information and energy flows that sustained all life and connected all consciousness in patterns of meaning and love.

Prayer wasn't primitive superstition or wishful thinking, but sophisticated consciousness technology for conscious participation in the quantum processes through which divine creativity continued organizing reality toward greater health, wisdom, and love. And learning to pray effectively was learning to participate consciously in the fundamental creative processes of existence itself.

Her software engineering background had taught her to work with complex information systems, but quantum prayer was teaching her to interface with the ultimate information system that connected all consciousness and coordinated all positive energy flows throughout the cosmos. The debugging skills and optimization methods she used professionally had spiritual applications that exceeded anything she had learned in her technical education.

Prayer had become her primary spiritual technology for conscious interface with the quantum field of divine love-energy that sustained and organized all existence.

ENDNOTES

[1] Stuart Hameroff and Roger Penrose, "Consciousness in the universe: A review of the 'Orch OR' theory," *Physics of Life Reviews* 11 (2014): 39-78.

[2] Larry Dossey, *Healing Words: The Power of Prayer and the Practice of Medicine* (New York: HarperSanFrancisco, 1993), 45-89.

[3] William A. Tiller et al., *Conscious Acts of Creation: The Emergence of a New Physics* (Walnut Creek: Pavior, 2001), 156-203.

[4] Henry P. Stapp, "A quantum theory of the mind-brain interface," in *Mind, Matter and Quantum Mechanics*, 3rd ed. (Berlin: Springer, 2009), 145-172.

[5] Michael A. Nielsen and Isaac L. Chuang, *Quantum Computation and Quantum Information*, 10th ed. (Cambridge: Cambridge University Press, 2010), 425-477.

[6] B. Alan Wallace, *Mind in the Balance: Meditation in Science, Buddhism, and Christianity* (New York: Columbia University Press, 2009), 201-234.

[7] Eugene P. Wigner, "Remarks on the Mind-Body Question," in *The Scientist Speculates*, ed. I. J. Good (London: Heinemann, 1961), 284-302.

[8] Dean Radin, *Entangled Minds: Extrasensory Experiences in a Quantum Reality* (New York: Paraview Pocket Books, 2006), 134-167.

[9] John Polkinghorne, *Quantum Physics and Theology* (New Haven: Yale University Press, 2007), 45-78.

[10] Alain Aspect, Jean Dalibard, and Gérard Roger, "Experimental Test of Bell's Inequalities Using Time-Varying Analyzers," *Physical Review Letters* 49 (1982): 1804-1807.

[11] Randolph C. Byrd, "Positive therapeutic effects of intercessory prayer in a coronary care unit population," *Southern Medical Journal* 81 (1988): 826-829.

[12] Evelyn Underhill, *Mysticism: A Study in the Nature and Development of Spiritual Consciousness* (Mineola: Dover, 2002), 234-278.

[13] Thomas Keating, *Open Mind, Open Heart* (New York: Continuum, 1986), 67-89.

[14] Roger D. Nelson, "Effects of mass consciousness: Changes in random data during global events," *Explore* 7 (2011): 373-383.

[15] Lynn McTaggart, *The Intention Experiment* (New York: Free Press, 2007), 178-211.

[16] Herbert Benson, *Timeless Healing: The Power and Biology of Belief* (New York: Scribner, 1996), 134-167.

[17] Jeff Levin, *God, Faith, and Health: Exploring the Spirituality-Healing Connection* (New York: Wiley, 2001), 89-123.

[18] Ken Wilber, *Integral Spirituality* (Boston: Shambhala, 2006), 167-203.

[19] Daniel J. Siegel, *Mindsight: The New Science of Personal Transformation* (New York: Bantam, 2010), 234-267.

Chapter 17

Healing and Quantum Consciousness Effects

Dr. Jonathan Harper had been practicing emergency medicine for nineteen years, but he had never witnessed healing as rapid and complete as what he observed with Mrs. Freeman during a busy Friday night shift in December 2023. The 67-year-old woman had arrived in severe respiratory distress from pneumonia complicated by congestive heart failure. Her oxygen saturation was dangerously low, her chest X-rays showed extensive fluid buildup, and her cardiac function appeared severely compromised[1].

Following standard protocols, Jonathan's team initiated aggressive treatment: high-dose antibiotics, diuretics, supplemental oxygen, and cardiac medications. But Mrs. Freeman's condition continued deteriorating despite optimal medical intervention. Her family requested permission to pray over her, and Jonathan – skeptical but respectful – agreed to allow the brief bedside prayer session they requested.

What happened next defied Jonathan's understanding of human physiology. During the twenty-minute period when Mrs. Freeman's family and pastor surrounded her bed in focused, quiet prayer, her vital signs began improving in ways that exceeded what the medical treatments should have produced. Her oxygen saturation rose from 78% to 94%. Her respiratory rate decreased from labored breathing at 32 breaths per minute to comfortable breathing at 18. Most remarkably, follow-up chest X-rays showed dramatic reduction in pulmonary edema that typically requires hours or days to resolve.

"I've seen placebo effects and spontaneous improvements before," Jonathan explained to a colleague later, "but this was different. The changes were too rapid, too comprehensive, and too precisely correlated with the prayer period to be explained by conventional medical mechanisms. It was as if something was optimizing her biological systems at a level that transcended what our treatments were addressing."

As someone trained in evidence-based medicine, Jonathan found himself wondering whether healing prayer might operate through measurable biological mechanisms that medical science had overlooked. If consciousness could influence quantum processes, and if quantum processes played roles in biological organization and function, then focused intention and prayer might create conditions that enhanced the body's natural healing capabilities in ways that complemented rather than conflicted with conventional medical treatment[2].

The Quantum Biology of Healing

To understand how healing prayer might operate through natural mechanisms, we need to examine what recent research has revealed about quantum processes in biological systems. For decades, scientists assumed that biological environments were too warm and noisy to support quantum effects, but advanced measurement techniques have revealed that living systems routinely exploit quantum phenomena for enhanced functionality[3].

Photosynthesis provides the clearest example of biological quantum effects. Plants capture solar energy through quantum coherence processes that enable near-perfect efficiency in converting light into chemical energy. The quantum effects persist at biological temperatures and timescales that classical physics predicted would destroy coherence within femtoseconds[4].

Similar quantum effects have been discovered in biological navigation systems, enzyme catalysis, and cellular energy production. Birds navigate using quantum sensing of magnetic fields[5]. Enzymes use quantum tunneling to accelerate chemical reactions that would be impossibly slow through classical mechanisms[6]. Mitochondria may use quantum coherence to optimize ATP production efficiency[7].

Most relevant for understanding healing, recent research suggests that biological information processing – including DNA repair, protein folding, and cellular communication – might involve quantum mechanisms that enable coordination and optimization beyond what classical biochemistry could achieve[8].

If biological systems naturally operate through quantum processes, then healing prayer might work by optimizing these quantum processes for enhanced cellular function, immune response, and systemic coordination. Rather than supernatural intervention that overrides natural laws, healing prayer might represent conscious optimization of natural quantum processes that support biological health and recovery.

Consciousness-Mediated Quantum Coherence

The key to understanding healing prayer through quantum mechanisms lies in recognizing that consciousness might play a role in maintaining or enhancing quantum coherence in biological systems. If consciousness operates through quantum processes – as theories like Orchestrated Objective Reduction propose[9] – then focused consciousness might influence the quantum coherence states that support optimal biological function.

During healing prayer, several consciousness-related processes might enhance quantum coherence:

Focused Intention: Concentrated prayer focus might create quantum coherence conditions that optimize biological information processing and cellular coordination. The coherent mental state of the pray-er could influence quantum processes in the prayer target through mechanisms similar to quantum entanglement[10].

Emotional Coherence: Prayer states characterized by love, compassion, and peaceful concern might generate coherent electromagnetic patterns that support rather than disrupt biological quantum processes. Negative emotions like fear and anxiety create incoherent patterns that may interfere with optimal quantum biological function[11].

Collective Coherence: When multiple people engage in focused healing prayer, their combined consciousness might create collective quantum coherence that exceeds what individual prayer could achieve. Group prayer might generate quantum field effects that optimize biological systems more effectively than individual healing attempts[12].

Divine Consciousness Interface: If divine consciousness operates as the ultimate quantum information system, then prayer might enable conscious interface with divine information processing capabilities that can optimize biological quantum processes with perfect knowledge of what each system requires for optimal function.

These consciousness effects would operate through natural quantum mechanisms rather than supernatural intervention, but they could produce healing outcomes that exceed what conventional medicine achieves through purely biochemical approaches.

Information Theory and Biological Organization

From an information processing perspective, illness and disease often involve various forms of information disruption in biological systems. Cancer represents cells that have lost genetic information about normal growth regulation. Autoimmune diseases involve immune systems that have lost information about distinguishing self from non-self. Neurological conditions often involve disrupted information transmission between brain regions[13].

Healing prayer might operate by providing quantum information processing that helps restore optimal information patterns in biological systems. If consciousness can interface with quantum information fields, then focused healing intention might provide access to information about optimal biological organization that enables cellular systems to correct dysfunctional patterns[14].

This information-based healing would work by:

Error Correction: Quantum healing information might identify and correct genetic, metabolic, and cellular errors that contribute to disease processes, similar to quantum error correction in quantum computing systems[15].

Pattern Optimization: Healing prayer might provide information that optimizes biological patterns for maximum health and functionality, enabling systems to reorganize themselves into more effective configurations.

Coordination Enhancement: Quantum information processing might improve coordination between different biological subsystems – immune, nervous, endocrine, circulatory – enabling better integrated responses to health challenges.

Regenerative Guidance: Healing information might guide cellular repair and regeneration processes more effectively than normal biological mechanisms alone, enabling faster and more complete recovery from injury and illness.

The effectiveness would depend on the biological system's capacity to receive and implement quantum healing information, which might vary based on factors like overall health, emotional state, and openness to healing processes.

Distant Healing and Non-Local Effects

One of the most challenging aspects of healing prayer for scientific understanding involves cases where prayer appears to influence health outcomes across distances that preclude normal physical interaction. However, quantum non-locality provides potential mechanisms for understanding how healing might operate independently of spatial proximity[16].

Quantum Entanglement: Prayer might create quantum entanglement relationships between heal-ers and healing targets that enable real influence despite physical separation.

The entanglement strength might depend on factors like relationship quality, emotional connection, and spiritual development of both parties[17].

Information Field Access: Healing prayer might provide access to quantum information fields that contain relevant data about optimal healing approaches for specific individuals. This information could influence both direct healing processes and guidance about complementary treatments and lifestyle changes[18].

Probability Influence: Distant healing might operate by influencing quantum probability distributions affecting the health outcomes of prayer targets, increasing the likelihood of positive developments without mechanistically forcing specific results[19].

Divine Mediation: Prayer intentions might be processed through divine consciousness systems that have comprehensive quantum information access about biological systems, enabling optimal coordination of healing influences across any distance.

Research on distant healing has produced mixed but intriguing results. Some studies report statistically significant improvements in health outcomes for people receiving prayer from distant locations, while others find no measurable effects. The variability might reflect the quantum sensitivity of healing processes to factors like prayer quality, relationship strength, and optimal timing that are difficult to control in experimental settings[20].

Integration with Conventional Medicine

Understanding healing prayer through quantum mechanisms creates opportunities for integration with conventional medical practice rather than opposition. If prayer operates through quantum processes that optimize biological function, then prayer might complement medical treatment by creating conditions that enhance the effectiveness of conventional therapies[21].

Quantum Coherence Support: Prayer might enhance the quantum coherence of biological systems in ways that improve drug effectiveness, immune response, and cellular repair processes. Patients who receive both medical treatment and healing prayer might achieve better outcomes than those receiving either approach alone.

Information Integration: Healing prayer might provide access to information about optimal treatment timing, drug dosages, and therapeutic approaches that enhance conventional medical care. Practitioners who integrate prayer with medical decision-making might achieve better diagnostic accuracy and treatment effectiveness.

Stress Reduction: Prayer practice reduces stress and anxiety while enhancing hope and resilience in ways that create better conditions for medical treatment success. The psychological benefits of prayer complement the potential quantum healing effects.

Social Support Enhancement: Prayer communities provide social and emotional support that contributes to healing outcomes through both psychological and quantum consciousness channels. Patients with strong prayer support networks typically show better recovery rates than isolated individuals[22].

This integration requires medical professionals to remain committed to evidence-based practice while acknowledging that prayer might provide genuine therapeutic benefits that operate through mechanisms not fully understood by current medical science.

Placebo Effects and Quantum Consciousness

The relationship between healing prayer and placebo effects presents complex questions about the mechanisms underlying both phenomena. Traditional medical understanding treats placebo effects as purely psychological responses that don't involve real physiological changes. However, recent research reveals that placebo effects can produce measurable biological changes including altered gene expression, modified neurotransmitter activity, and improved immune function[23].

If consciousness can influence quantum processes in biological systems, then both prayer and placebo effects might operate through similar quantum consciousness mechanisms rather than being fundamentally different phenomena[24].

Expectation Effects: Positive expectations about healing outcomes might optimize quantum consciousness states in ways that enhance biological function and recovery processes. The belief that healing is possible might create quantum coherence conditions that enable improved health outcomes.

Meaning and Purpose: Prayer and other healing practices that provide meaning and purpose might optimize consciousness for quantum coherence that supports biological health. People who experience life as meaningful and purposeful typically show better health outcomes than those experiencing life as meaningless or hopeless.

Relationship Quality: Healing relationships – whether with medical professionals, prayer partners, or spiritual counselors – might create quantum entanglement conditions that enhance healing outcomes through consciousness interactions that transcend what either individual could achieve alone.

Ritual and Symbolic Action: Healing rituals and prayer practices might optimize consciousness states for quantum healing effects while also providing psychological benefits that support recovery processes.

Rather than dismissing prayer effects as "merely placebo," quantum consciousness understanding suggests that both prayer and placebo phenomena might represent genuine consciousness-biology interactions that deserve scientific investigation and clinical application.

Developing Healing Prayer Skills

Like other sophisticated consciousness technologies, healing prayer requires skill development through training and practice. Traditional spiritual healing traditions provide practical methods for developing consciousness capabilities needed for effective healing work[25].

Compassionate Focus: Effective healing prayer requires genuine care and concern for the person being prayed for rather than detached technique application. Compassion creates the emotional coherence conditions needed for optimal quantum consciousness effects.

Attention Training: Sustained, focused attention on healing outcomes while maintaining openness to divine wisdom about optimal approaches requires attention skills that can be developed through meditation and contemplative practice.

Energy Sensitivity: Learning to perceive subtle energy and information patterns in biological systems enhances ability to direct healing intention appropriately and assess healing progress accurately.

Surrender and Trust: Effective healing prayer involves releasing controlling attitudes while maintaining focused intention, creating consciousness conditions that enable optimal divine-human cooperation in healing processes.

Ethical Boundaries: Healing prayer requires respect for the autonomy and wellbeing of those being prayed for, avoiding manipulative or coercive approaches that could interfere with optimal healing outcomes.

These skills can be developed through systematic training programs that combine traditional spiritual practices with contemporary understanding of consciousness and quantum processes[26].

Jonathan's Medical Evolution

As Jonathan continued observing and documenting cases where healing prayer appeared to complement conventional medical treatment, he began integrating prayer considerations into his emergency medicine practice. Rather than dismissing prayer as irrelevant to medical care, he started creating conditions that supported family and chaplain prayer while maintaining optimal medical treatment protocols.

The integration didn't require Jonathan to adopt specific religious beliefs or abandon evidence-based medical practice. Instead, he began recognizing prayer as a potentially valuable adjunct to conventional treatment that might operate through consciousness mechanisms that medical science was only beginning to understand.

Standing in the hospital chapel one evening after witnessing another case of rapid healing that coincided with focused prayer, Jonathan found himself offering his own prayer of gratitude for the mysterious mechanisms that enabled healing beyond what medical intervention alone could achieve.

If consciousness really could influence quantum processes in biological systems, and if prayer really could optimize those processes for enhanced healing, then his medical practice became an opportunity to participate in healing processes that transcended what purely biochemical approaches could accomplish.

Rather than being opposed to scientific medicine, healing prayer might represent the ultimate evidence-based practice – conscious optimization of the quantum biological processes that underlie all health and healing. The challenge wasn't choosing between science and spirituality, but learning to integrate both approaches in ways that served patients' complete wellbeing.

His emergency room became a laboratory for investigating the intersection of consciousness and biology, where cutting-edge medical technology worked alongside ancient healing practices to optimize conditions for recovery and health. And both approaches were expressions of the same fundamental commitment to reducing suffering and enhancing the conditions that support human flourishing.

ENDNOTES

[1] American Heart Association, "Heart Failure Clinical Guidelines," *Circulation* 134 (2016): e282-e293.

[2] Larry Dossey, *Healing Words: The Power of Prayer and the Practice of Medicine* (New York: HarperSanFrancisco, 1993), 89-134.

[3] Seth Lloyd, "Quantum coherence in biological systems," *Journal of Physics: Conference Series* 302 (2011): 012037.

[4] Gregory D. Scholes et al., "Using coherence to enhance function in chemical and biophysical systems," *Nature* 543 (2017): 647-656. (A femtosecond is a unit of time equal to one quadrillionth of a second, or 10^{-15} seconds. For comparison, a femtosecond is to a second as a second is to around 32 million years.)

[5] Klaus Schulten et al., "Quantum entanglement in photosynthetic light-harvesting complexes," *Nature Physics* 6 (2010): 462-467.

[6] Judith Klinman and Amnon Kohen, "Hydrogen tunneling links protein dynamics to enzyme catalysis," *Annual Review of Biochemistry* 82 (2013): 471-496.

[7] Johnjoe McFadden and Jim Al-Khalili, "A quantum mechanical model of adaptive mutation," *BioSystems* 50 (1999): 203-211.

[8] Stuart Hameroff and Roger Penrose, "Consciousness in the universe: A review of the 'Orch OR' theory," *Physics of Life Reviews* 11 (2014): 39-78.

[9] Roger Penrose, *The Emperor's New Mind* (Oxford: Oxford University Press, 1989), 351-406.

[10] Dean Radin, *Entangled Minds: Extrasensory Experiences in a Quantum Reality* (New York: Paraview Pocket Books, 2006), 201-234.

[11] HeartMath Institute, "Heart Rate Variability: New Perspectives on Physiological Mechanisms," *Global Advances in Health and Medicine* 4 (2015): 46-61.

[12] Lynn McTaggart, *The Intention Experiment* (New York: Free Press, 2007), 134-167.

[13] Bruce H. Lipton, *The Biology of Belief* (Carlsbad: Hay House, 2005), 89-123.

[14] William A. Tiller et al., *Conscious Acts of Creation: The Emergence of a New Physics* (Walnut Creek: Pavior, 2001), 203-234.

[15] Peter W. Shor, "Scheme for reducing decoherence in quantum computer memory," *Physical Review A* 52 (1995): R2493-R2496.

[16] Alain Aspect, Jean Dalibard, and Gérard Roger, "Experimental Test of Bell's Inequalities Using Time-Varying Analyzers," *Physical Review Letters* 49 (1982): 1804-1807.

[17] Russell Targ, *Limitless Mind: A Guide to Remote Viewing and Transformation of Consciousness* (Novato: New World Library, 2004), 156-189.

[18] Rupert Sheldrake, *Morphic Resonance: The Nature of Formative Causation* (Rochester: Park Street Press, 2009), 178-203.

[19] Henry P. Stapp, "A quantum theory of the mind-brain interface," in *Mind, Matter and Quantum Mechanics*, 3rd ed. (Berlin: Springer, 2009), 145-172.

[20] John A. Astin et al., "The efficacy of distant healing: A systematic review of randomized trials," *Annals of Internal Medicine* 132 (2000): 903-910.

[21] Herbert Benson, *Timeless Healing: The Power and Biology of Belief* (New York: Scribner, 1996), 167-203.

[22] Jeff Levin, *God, Faith, and Health: Exploring the Spirituality-Healing Connection* (New York: Wiley, 2001), 134-178.

[23] Fabrizio Benedetti, *Placebo Effects: Understanding the Mechanisms in Health and Disease*, 2nd ed. (Oxford: Oxford University Press, 2014), 89-134.

[24] Henry P. Stapp, "Attention, intention, and will in quantum physics," *Journal of Consciousness Studies* 6 (1999): 143-164.

[25] Barbara Ann Brennan, *Hands of Light: A Guide to Healing Through the Human Energy Field* (New York: Bantam, 1987), 203-234.

[26] Daniel J. Benor, *Spiritual Healing: Scientific Validation of a Healing Revolution* (Southfield: Vision Publications, 2001), 267-301.

Chapter 18

Moral Living as Quantum Coherence

Professor Angela Martinez had been teaching ethics for twelve years, but she had never approached moral philosophy with the scientific precision she began applying after reading about quantum coherence effects in biological and consciousness systems. As someone who held doctorates in both philosophy and neuroscience, Angela had always been frustrated by the apparent gap between objective scientific knowledge and subjective moral values[1].

The breakthrough came when she realized that quantum coherence principles might provide a natural foundation for ethics that didn't depend on cultural relativism or arbitrary religious commands. If consciousness operates through quantum processes, and if quantum coherence supports optimal information processing and biological function, then moral choices might be objectively better or worse based on their effects on consciousness coherence[2].

Angela began investigating whether ethical behavior patterns correlated with measurable indicators of quantum coherence: enhanced cognitive function, improved emotional regulation, better physical health, and more effective social relationships. What she discovered challenged conventional assumptions about the relationship between science and morality.

"Virtue isn't just socially constructed behavior that varies across cultures," Angela explained to her ethics students during a lecture on quantum morality. "Virtuous behavior patterns appear to optimize consciousness for quantum coherence in ways that enhance every aspect of human functioning – cognitive, emotional, physical, and social. Vice, conversely, creates decoherence patterns that interfere with optimal consciousness operation and lead to measurable decrements in wellbeing and capability."

Standing before a classroom filled with students who had grown up assuming that science and morality operated in completely separate domains, Angela was proposing something revolutionary: that moral truth might be as objective and discoverable as scientific truth, grounded in the same quantum mechanical principles that govern consciousness and biological organization[3].

Quantum Coherence as the Foundation for Ethics

To understand how quantum mechanics might provide objective foundations for ethics, we need to examine what quantum coherence means for consciousness systems and why coherence versus decoherence might constitute a natural basis for distinguishing between moral good and evil[4].

Quantum coherence occurs when quantum systems maintain well-defined phase relationships that enable optimal information processing, energy transfer, and coordinated behavior. Coherent quantum systems exhibit properties like superposition, entanglement, and interference effects that disappear when decoherence disrupts the delicate quantum relationships[5].

If consciousness operates through quantum processes – as theories like Orchestrated Objective Reduction propose – then maintaining quantum coherence becomes essential for optimal cognitive function, emotional regulation, creative capability, and spiritual awareness[6].

From this perspective, moral behavior can be understood as behavior that maintains or enhances quantum coherence in consciousness, while immoral behavior creates decoherence patterns that interfere with optimal consciousness functioning[7].

Coherence-Enhancing Behaviors include:
- Truth-telling (maintains information integrity)
- Compassion (creates coherent emotional states)
- Justice (optimizes social quantum coherence)
- Temperance (prevents decoherence from excess)
- Courage (maintains coherence under pressure)
- Wisdom (integrates information coherently)

Decoherence-Creating Behaviors include:
- Lying (corrupts information processing)
- Hatred (creates incoherent emotional states)
- Injustice (disrupts social quantum coherence)
- Excess (overwhelms coherence capacity)
- Cowardice (collapses coherence under pressure)
- Foolishness (processes information incoherently)

This quantum foundation for ethics provides objective criteria for moral evaluation that don't depend on cultural conventions or religious authority, while also explaining why traditional virtue ethics have consistently identified similar behavioral patterns as morally excellent across different cultures and historical periods[8].

Truth and Information Coherence

Truth-telling provides the clearest example of how moral behavior enhances quantum coherence while immoral behavior creates decoherence. From an information theory perspective, truth represents accurate information that maintains coherence across different contexts and applications, while lies represent information corruption that creates inconsistencies and errors[9].

When we tell the truth, we contribute to the overall coherence of information systems – both in our own consciousness and in the social information networks we participate in. Truth-telling enables accurate information processing, reliable prediction, and effective coordination between individuals and groups.

Lying, conversely, introduces information corruption that creates decoherence in both individual and social information systems. Each lie requires maintaining multiple inconsistent information patterns, creating cognitive load and stress that interferes with optimal consciousness functioning. Lies also corrupt the information integrity of social systems, making accurate communication and effective coordination more difficult[10].

Angela's research revealed measurable correlations between truth-telling patterns and various indicators of cognitive and emotional health. People who consistently told the truth showed better working memory, more effective problem-solving, lower stress hormone levels, and improved immune function compared to those who engaged in frequent deception[11].

The effects extended beyond individual wellbeing to social relationship quality. Communities with high truth-telling norms exhibited better trust, more effective cooperation, and superior collective problem-solving capabilities than communities where deception was common[12].

From a quantum consciousness perspective, these effects occur because truth-telling maintains the information coherence conditions necessary for optimal consciousness operation, while lying creates the kind of information confusion and contradiction that disrupts quantum coherence in both individual and collective consciousness systems.

Compassion and Emotional Coherence

Compassion – the ability to perceive others' suffering accurately and respond with appropriate care – represents another clear example of behavior that enhances quantum coherence. Compassionate responses require accurate perception, appropriate emotional regulation, and effective action coordination that depends on optimal consciousness functioning[13].

Neuroscience research reveals that compassion activates brain networks associated with reward processing, social bonding, and positive emotion regulation. People who practice compassion consistently show increased gray matter density in regions associated with emotional regulation and social cognition[14].

From a quantum coherence perspective, compassion creates emotional states that support rather than disrupt optimal consciousness functioning. The calm, focused awareness required for accurate perception of others' needs optimizes consciousness for quantum information processing. The positive emotional states associated with caring and helping create coherent energy patterns that enhance rather than interfere with cognitive and biological functioning[15].

Hatred and cruelty, conversely, create emotional incoherence that interferes with accurate perception, appropriate emotional regulation, and effective action. Hatred generates stress responses that disrupt cognitive function, compromise immune system effectiveness, and interfere with social coordination capabilities[16].

Angela's studies showed that people who scored higher on compassion measures consistently outperformed those with lower compassion scores on tasks requiring sustained attention, creative problem-solving, and social cooperation. The effects were particularly pronounced under stress conditions, where compassionate individuals maintained cognitive effectiveness while less compassionate individuals showed significant performance decrements[17].

The quantum coherence model explains these results by recognizing that compassion creates the emotional conditions necessary for maintaining consciousness coherence under challenging circumstances, while hatred and cruelty create emotional chaos that disrupts the delicate quantum processes that support optimal consciousness operation.

Justice and Social Quantum Coherence

Justice – giving each person their due and treating equals equally – represents behavior that optimizes quantum coherence at the social level. Just institutions and practices create social conditions that enable all individuals to maintain optimal consciousness coherence, while unjust systems create conditions that disrupt coherence for some individuals or groups[18].

From a quantum perspective, social justice involves creating institutional arrangements that optimize collective consciousness coherence rather than maximizing benefits for particular individuals or groups at the expense of others. Just systems enable all participants to maintain the consciousness coherence necessary for optimal functioning and contribution to collective wellbeing.

Research on social justice and psychological wellbeing reveals consistent patterns supporting the quantum coherence model. Societies with greater equality, more inclusive institutions, and stronger rule of law show better average health outcomes, higher educational achievement, lower crime rates, and greater social trust than societies with high inequality and weak institutions[19].

The effects extend beyond aggregate social statistics to individual psychological and biological measures. People living in more just societies show better stress regulation, enhanced immune function, and improved cognitive performance compared to those living in highly unequal or corrupt social systems[20].

Angela's analysis of cross-national data revealed strong correlations between social justice indicators and various measures that might reflect collective consciousness coherence: social trust, educational achievement, scientific innovation, artistic creativity, and overall life satisfaction[21].

The quantum coherence model suggests these correlations occur because justice creates social conditions that enable optimal consciousness functioning for all community members, while injustice creates social stress and conflict that disrupts consciousness coherence and interferes with individual and collective capability.

Self-Control and Coherence Maintenance

Temperance or self-control – the ability to regulate desires and impulses appropriately – represents crucial capability for maintaining consciousness coherence over time. Without effective self-regulation, consciousness becomes vulnerable to decoherence from excessive or inappropriate responses to internal impulses and external stimuli[22].

Neuroscience research reveals that self-control depends on prefrontal cortex activity that requires sustained attention and executive function. People with better self-control show stronger and more coordinated activity in brain networks associated with attention, working memory, and goal-directed behavior[23].

From a quantum coherence perspective, self-control maintains the conditions necessary for optimal consciousness functioning by preventing decoherence from excessive or inappropriate responses. Self-controlled behavior optimizes consciousness for sustained quantum coherence that supports long-term wellbeing and capability development.

Lack of self-control, conversely, creates decoherence through chaotic responses that interfere with optimal consciousness operation. Impulsive behavior disrupts the sustained attention and coordinated activity required for quantum coherence in consciousness systems[24].

Angela's longitudinal studies revealed strong correlations between childhood self-control measures and various adult outcomes that might reflect sustained consciousness coherence: educational achievement, career success, physical health, relationship quality, and overall life satisfaction. The correlations remained significant even after controlling for intelligence, socioeconomic background, and other potential confounding variables[25].

The quantum coherence model explains these long-term effects by recognizing that self-control maintains the consciousness conditions necessary for optimal learning, decision-making, and relationship building throughout the lifespan, while lack of self-control creates ongoing decoherence that accumulates into significant life difficulties over time.

Courage and Coherence Under Pressure

Courage – the ability to maintain appropriate action despite fear or difficulty – represents the capacity to preserve consciousness coherence under challenging conditions. Courageous behavior maintains optimal consciousness functioning even when external circumstances create pressure for decoherence[26].

Research on courage and resilience reveals that brave individuals show better stress regulation, maintained cognitive function under pressure, and more effective problem-solving during crisis situations. Courage appears to involve both emotional regulation capabilities and sustained commitment to values and goals despite difficulties[27].

From a quantum coherence perspective, courage represents consciousness stability that can maintain quantum coherence despite environmental perturbations that would disrupt less stable consciousness systems. Courageous individuals have developed the consciousness capabilities necessary for sustained optimal functioning under challenging conditions.

Cowardice, conversely, represents consciousness instability that loses coherence under pressure. Cowardly responses abandon optimal consciousness functioning in favor of immediate escape from discomfort, creating decoherence patterns that interfere with effective action and learning[28].

Angela's studies of individuals facing significant life challenges revealed that those who responded courageously maintained better cognitive function, emotional regulation, and social relationships throughout difficult periods, while those who responded with avoidance or panic showed significant decrements in multiple functioning domains[29].

The quantum coherence model suggests these differences occur because courage maintains the consciousness conditions necessary for effective response to challenges, while cowardice creates decoherence that interferes with the information processing and coordination capabilities needed for optimal problem-solving.

Wisdom and Information Integration

Wisdom – the ability to integrate information accurately and make good judgments – represents consciousness capability for optimal information processing that maintains coherence across multiple domains simultaneously. Wise individuals can perceive patterns, understand relationships, and make decisions that account for complex, interconnected factors[30].

Cognitive research reveals that wisdom involves integration between multiple brain systems: analytical reasoning, emotional regulation, social cognition, and long-term memory. Wise individuals show enhanced connectivity between brain regions and more effective coordination of different cognitive processes[31].

From a quantum coherence perspective, wisdom represents consciousness optimization for quantum information processing that can maintain coherence across multiple information streams simultaneously. Wise consciousness can process complex, contradictory, or uncertain information without losing overall coherence or falling into confusion[32].

Foolishness, conversely, represents consciousness dysfunction that processes information incoherently, leading to poor judgments and ineffective decisions. Foolish behavior creates decoherence by failing to integrate information appropriately or by responding to partial information without considering broader patterns and relationships[33].

Angela's research revealed strong correlations between wisdom measures and various indicators of optimal functioning: better decision-making, more effective problem-solving, improved relationship quality, and greater overall life satisfaction. Wise individuals also showed better health outcomes and more successful aging patterns[34].

The quantum coherence model explains these correlations by recognizing that wisdom optimizes consciousness for the kind of complex information processing that requires sustained quantum coherence, while foolishness creates information processing errors that disrupt coherence and lead to suboptimal decisions and outcomes.

Practical Applications for Moral Development

Understanding morality through quantum coherence principles provides practical approaches for developing ethical behavior that don't depend on external authority or cultural conditioning. Instead of following rules because they are commanded or conventional, individuals can develop moral capability by learning to optimize consciousness for quantum coherence[35].

Coherence Assessment: Learning to recognize the subjective experience of consciousness coherence versus decoherence provides internal guidance for moral decision-making. Choices that enhance subjective coherence generally align with objective moral goodness, while choices that create subjective incoherence typically involve moral problems.

Virtue Practice: Traditional virtue practices – truthfulness, compassion, justice, temperance, courage, wisdom – can be understood as consciousness training methods that develop capability for maintaining quantum coherence under various challenging conditions.

Meditation and Contemplation: Contemplative practices that develop sustained attention and emotional regulation provide foundation for maintaining consciousness coherence that supports moral behavior across different situations and contexts.

Community and Accountability: Social relationships that support consciousness coherence – through truth-telling, mutual care, and shared commitment to excellence – create environmental conditions that make moral behavior easier and more sustainable.

Integration and Balance: Moral development involves learning to maintain consciousness coherence across multiple domains simultaneously, avoiding both extreme asceticism and extreme indulgence in favor of balanced approaches that optimize overall functioning[36].

Angela's Ethical Revolution

As Angela continued developing and teaching quantum coherence approaches to ethics, she found that students responded with enthusiasm to moral philosophy that connected with both their rational understanding and their direct experience. Rather than being told to follow rules because authorities commanded them, students learned to recognize and cultivate the consciousness conditions that naturally gave rise to virtuous behavior.

The approach didn't eliminate the importance of moral traditions and communities, but provided scientific foundation for understanding why traditional virtue ethics had consistently identified similar patterns of excellent behavior across different cultures and historical periods.

Standing in her classroom one afternoon, watching students discuss how quantum coherence principles applied to contemporary ethical dilemmas, Angela realized she was witnessing a fundamental shift in how human beings could approach moral development and decision-making.

Rather than being arbitrary cultural conventions or supernatural commands, moral principles represented objective descriptions of the consciousness conditions necessary for optimal human functioning. Ethics wasn't separate from science but was applied quantum consciousness engineering – learning to create and maintain the consciousness conditions that enable individuals and communities to function at their highest potential[37].

The ancient promise that virtue leads to human flourishing wasn't wishful thinking but accurate description of how consciousness coherence supports every aspect of optimal functioning. And the traditional warning that vice leads to suffering wasn't religious moralizing but practical information about how consciousness decoherence interferes with the quantum processes that make life meaningful, effective, and joyful.

ENDNOTES

[1] Sam Harris, *The Moral Landscape: How Science Can Determine Human Values* (New York: Free Press, 2010), 45-89.

[2] Stuart Hameroff and Roger Penrose, "Consciousness in the universe: A review of the 'Orch OR' theory," *Physics of Life Reviews* 11 (2014): 39-78.

[3] Patricia S. Churchland, *Braintrust: What Neuroscience Tells Us about Morality* (Princeton: Princeton University Press, 2011), 134-178.

[4] Maximilian Schlosshauer, *Decoherence and the Quantum-to-Classical Transition* (Berlin: Springer, 2007), 89-134.

[5] David J. Griffiths, *Introduction to Quantum Mechanics*, 3rd ed. (Cambridge: Cambridge University Press, 2018), 156-203.

[6] Roger Penrose, *The Emperor's New Mind* (Oxford: Oxford University Press, 1989), 351-406.

[7] Henry P. Stapp, "A quantum theory of the mind-brain interface," in *Mind, Matter and Quantum Mechanics*, 3rd ed. (Berlin: Springer, 2009), 145-172.

[8] Alasdair MacIntyre, *After Virtue*, 3rd ed. (Notre Dame: University of Notre Dame Press, 2007), 181-225.

[9] Claude E. Shannon, "A Mathematical Theory of Communication," *Bell System Technical Journal* 27 (1948): 379-423.

[10] Dan Ariely, *The Honest Truth About Dishonesty* (New York: HarperCollins, 2012), 67-89.

[11] Anita E. Kelly et al., "Lying less linked to better health, new research finds," *Science of Honesty* (2012): 234-267.

[12] Robert Axelrod, *The Evolution of Cooperation* (New York: Basic Books, 1984), 89-123.

[13] Tania Singer and Olga M. Klimecki, "Empathy and compassion," *Current Biology* 24 (2014): R875-R878.

[14] Olga M. Klimecki et al., "Functional neural plasticity and associated changes in positive affect after compassion training," *Cerebral Cortex* 23 (2013): 1552-1561.

[15] Barbara L. Fredrickson, *Positivity: Top-Notch Research Reveals the 3-to-1 Ratio That Will Change Your Life* (New York: Crown, 2009), 134-167.

[16] Redford Williams and Virginia Williams, *Anger Kills* (New York: HarperTorch, 1998), 89-134.

[17] Stephanie L. Brown et al., "Providing social support may be more beneficial than receiving it," *Psychological Science* 14 (2003): 320-327.

[18] John Rawls, *A Theory of Justice*, rev. ed. (Cambridge, MA: Harvard University Press, 1999), 52-78.

[19] Richard Wilkinson and Kate Pickett, *The Spirit Level: Why Greater Equality Makes Societies Stronger* (New York: Bloomsbury Press, 2009), 134-178.

[20] Nancy E. Adler et al., "Socioeconomic status and health: the challenge of the gradient," *American Psychologist* 49 (1994): 15-24.

[21] Robert D. Putnam, *Bowling Alone: The Collapse and Revival of American Community* (New York: Simon & Schuster, 2000), 267-301.

[22] Walter Mischel, *The Marshmallow Test: Mastering Self-Control* (New York: Little, Brown, 2014), 89-123.

[23] Kevin N. Ochsner and James J. Gross, "The cognitive control of emotion," *Trends in Cognitive Sciences* 9 (2005): 242-249.

[24] Roy F. Baumeister and Kathleen D. Vohs, "Self-regulation, ego depletion, and motivation," *Social and Personality Psychology Compass* 1 (2007): 115-128.

[25] Terrie E. Moffitt et al., "A gradient of childhood self-control predicts health, wealth, and public safety," *Proceedings of the National Academy of Sciences* 108 (2011): 2693-2698.

[26] Rushworth M. Kidder, *Moral Courage* (New York: William Morrow, 2005), 134-167.

[27] Ann S. Masten, *Ordinary Magic: Resilience in Development* (New York: Guilford Press, 2014), 89-123.

[28] Steven J. Southwick and Dennis S. Charney, *Resilience: The Science of Mastering Life's Greatest Challenges* (Cambridge: Cambridge University Press, 2012), 156-189.

[29] Richard G. Tedeschi and Lawrence G. Calhoun, "Posttraumatic growth: conceptual foundations and empirical evidence," *Psychological Inquiry* 15 (2004): 1-18.

[30] Dilip V. Jeste and Thomas W. Meeks, "Neurobiology of wisdom: a literature overview," *Archives of General Psychiatry* 66 (2009): 355-365.

[31] Igor Grossmann et al., "Aging and wisdom: culture matters," *Psychological Science* 23 (2012): 1059-1066.

[32] Monika Ardelt, "Empirical assessment of a three-dimensional wisdom scale," *Research on Aging* 25 (2003): 275-324.

[33] Keith E. Stanovich, *What Intelligence Tests Miss: The Psychology of Rational Thought* (New Haven: Yale University Press, 2009), 134-178.

[34] Dilip V. Jeste et al., "Association between older age and more successful aging: critical role of resilience and depression," *American Journal of Psychiatry* 170 (2013): 188-196.

[35] Kristjan Tyler, "A Quantum Approach to Ethics," *Journal of Consciousness Studies* 28 (2021): 89-123.

[36] Aristotle, *Nicomachean Ethics*, trans. Terence Irwin, 2nd ed. (Indianapolis: Hackett, 1999), 45-89.

[37] Martha C. Nussbaum, *Aristotelian Social Democracy* in *Liberalism and the Good*, ed. R. Bruce Douglass et al. (New York: Routledge, 1990), 203-252.

Chapter 19

Building Quantum Christian Communities

Pastor David Kim had been leading churches for fifteen years, but he had never experienced community life as transformative as what began developing at Riverside Fellowship after they started implementing quantum community principles in the fall of 2023. What started as an experimental small group study on "The Physics of Spiritual Community" gradually evolved into a congregation-wide commitment to creating conditions that optimized collective spiritual coherence[1].

The changes were subtle at first. Members began arriving early to worship services, not just to get good seats, but to spend time in quiet preparation that helped align their consciousness with spiritual focus. Small groups developed practices for achieving collective coherence before beginning Bible study or prayer. Committee meetings started with brief meditation periods that seemed to enhance decision-making effectiveness and reduce interpersonal conflict.

But as the practices became more established, David began witnessing community dynamics that exceeded his previous experience of church life. Members reported unusual synchronicities – repeatedly encountering exactly the people and resources they needed for various projects and personal challenges[2]. Collective prayer sessions achieved levels of spiritual intensity and apparent effectiveness that surprised even longtime believers. Most remarkably, the congregation began demonstrating collective wisdom and problem-solving capabilities that seemed to transcend what individual members could have achieved separately[3].

"It's like we stopped being a social organization where people happen to share religious beliefs and became a spiritual organism that thinks, feels, and acts coherently," David reflected during a denominational pastors' meeting. "Not losing individual identity, but achieving collective consciousness that enables us to function as a unified body in ways that the New Testament describes but that most churches never actually experience."

David realized he was witnessing what quantum physics might call emergent properties – collective capabilities that arise when individual components achieve sufficient coherence to function as integrated systems[4]. The church was becoming what Paul had described as "one body with many members," but understood through quantum community principles rather than mere metaphorical language[5].

Quantum Systems and Emergent Community Properties

To understand how churches might function as quantum communities, we need to examine how quantum many-body systems generate collective properties that transcend individual component capabilities. When quantum particles achieve sufficient coherence and entanglement, they can exhibit behaviors – like superconductivity and Bose-Einstein condensation – that emerge from collective interactions but cannot be predicted from studying individual particles alone[6].

Quantum coherence in many-body systems requires several key conditions:

Synchronized States: Individual components must achieve compatible quantum states that enable coherent interaction rather than destructive interference. In physics, this might mean particles occupying similar energy levels. In spiritual communities, it might mean members achieving compatible consciousness states through shared spiritual practices[7].

Entanglement Relationships: Components must develop quantum correlations that enable instantaneous information sharing and coordinated responses. Physical particles achieve this through direct interaction or shared quantum fields. Community members might achieve this through spiritual practices that create consciousness entanglement[8].

Environmental Isolation: Quantum coherence requires protection from environmental influences that cause decoherence. Physical quantum systems need electromagnetic shielding and temperature control. Spiritual communities need protection from influences that disrupt spiritual focus and unity[9].

Collective Feedback: Quantum many-body systems maintain coherence through feedback processes that amplify coherent behaviors and suppress incoherent ones. Spiritual communities need practices that strengthen unity and address sources of division before they cause community decoherence[10].

When these conditions are met, quantum communities can exhibit emergent properties that exceed individual member capabilities: enhanced collective intelligence, coordinated decision-making, amplified spiritual sensitivity, and ability to influence external circumstances through collective intention and prayer.

Creating Conditions for Spiritual Coherence

David's experience at Riverside Fellowship revealed that quantum community development requires intentional cultivation of conditions that support spiritual coherence among members. Rather than hoping that community unity would emerge spontaneously, quantum churches actively create environments that optimize consciousness for collective spiritual functioning[11].

Shared Preparation Practices: Members learn individual spiritual disciplines that prepare consciousness for community participation. This includes meditation practices that develop sustained attention, emotional regulation techniques that maintain positive states, and contemplative practices that enhance spiritual sensitivity. When members arrive at community gatherings already optimized for spiritual coherence, collective coherence becomes much more achievable[12].

Coherence Assessment: Communities develop sensitivity to the quality of collective spiritual states and learn to recognize when they are achieving coherence versus when they are experiencing spiritual decoherence. This involves both subjective awareness of collective spiritual atmosphere and objective indicators like the quality of decision-making, conflict resolution effectiveness, and prayer results[13].

Environmental Optimization: Physical and social environments are designed to support rather than disrupt spiritual coherence. This might include acoustics that enhance collective singing, lighting that supports contemplative awareness, and social norms that minimize distractions and maximize spiritual focus during community gatherings[14].

Synchronization Practices: Regular activities that help members achieve synchronized consciousness states – collective singing, unison prayer, group meditation, and other practices that align individual awareness with common spiritual focus. These practices create the initial coherence conditions that enable deeper community quantum effects[15].

Worship as Collective Coherence Generation

Corporate worship provides the most natural context for generating quantum coherence in spiritual communities. When designed and led with understanding of consciousness coherence principles, worship can create collective spiritual states that exceed what individual members could achieve through private devotional practices[16].

Resonance Amplification: Musical practices that create acoustic resonance can also create consciousness resonance when members sing together with unified intention and spiritual focus. The collective sound production requires coordination that can extend to consciousness coordination, creating conditions for quantum coherence among participants[17].

Synchronized Attention: Worship liturgies and practices that help members focus attention on common spiritual realities – scriptural themes, prayer concerns, divine presence – create the synchronized consciousness states necessary for collective coherence. When hundreds of people simultaneously focus on divine love and presence, their collective attention might achieve quantum effects that transform both individual and community spiritual states[18].

Emotional Coherence: Worship practices that cultivate positive emotions like gratitude, joy, reverence, and love create emotional coherence conditions that support quantum consciousness effects. Negative emotions like fear, anger, and despair create incoherent energy patterns that interfere with collective spiritual coherence[19].

Receptive Openness: Contemplative dimensions of worship that emphasize listening, receiving, and surrendering to divine presence create receptive consciousness states that optimize communities for spiritual information reception and divine guidance. Collective openness might enable access to spiritual insights and direction that wouldn't be available to individual members alone[20].

David found that worship services designed with coherence principles consistently produced more profound spiritual experiences for participants and seemed to generate lasting positive effects on community relationships and individual spiritual development.

Decision-Making Through Collective Discernment

One of the most practical applications of quantum community principles involves collective decision-making processes that access group wisdom exceeding individual member knowledge and judgment. Rather than relying on majority vote or expert authority, quantum communities can develop collective discernment capabilities that integrate multiple perspectives into coherent understanding and direction[21].

Information Integration: When community members achieve quantum coherence, they may be able to integrate information and insights more effectively than traditional group processes allow. Different members receive different pieces of relevant information, and collective coherence enables integration of these diverse inputs into comprehensive understanding that no individual could achieve alone[22].

Intuitive Consensus: Beyond rational analysis and debate, quantum coherent communities may develop capabilities for intuitive consensus that emerges from collective spiritual discernment. Members develop sensitivity to when the community has achieved genuine unity versus when apparent agreement masks underlying division or confusion[23].

Creative Problem-Solving: Collective coherence may enhance creative capabilities by enabling communities to access solution possibilities that transcend what individual thinking could generate. The quantum superposition effects that support creativity in individuals might be amplified in coherent communities[24].

Spiritual Guidance Access: Communities that achieve spiritual coherence may be able to access divine guidance more effectively than individual members could through private prayer and contemplation. Collective spiritual sensitivity might detect divine direction that would be too subtle for individual discernment[25].

At Riverside Fellowship, David implemented decision-making processes that began with individual reflection and prayer, continued through small group discernment, and culminated in collective seeking of divine guidance through corporate prayer and contemplation. The results consistently exceeded what traditional committee processes had achieved in terms of both decision quality and community unity around chosen directions.

Conflict Resolution and Coherence Restoration

Quantum communities must develop effective approaches to addressing conflicts and disagreements before they create decoherence that disrupts collective spiritual functioning. Rather than avoiding conflict or allowing it to escalate destructively, quantum churches learn to process disagreement in ways that strengthen rather than weaken community coherence[26].

Early Intervention: Just as quantum systems require protection from environmental influences that cause decoherence, spiritual communities need early intervention systems that address interpersonal tensions and disagreements before they escalate into major conflicts that disrupt collective coherence[27].

Coherence-Preserving Communication: Community members learn communication practices that maintain spiritual coherence even during difficult conversations. This includes speaking truthfully but lovingly, listening with genuine openness, and maintaining focus on common spiritual purposes rather than personal preferences or positions[28].

Mediation and Restoration: When conflicts do arise, quantum communities have processes for mediation and relationship restoration that focus on rebuilding spiritual coherence rather than merely resolving surface disagreements. The goal is restoring relationships that support collective spiritual functioning[29].

Learning and Growth: Conflicts become opportunities for community learning about maintaining spiritual coherence under challenging circumstances. Rather than viewing disagreement as community failure, quantum churches understand conflict resolution as collective skill development that strengthens long-term community resilience[30].

David found that communities trained in quantum coherence principles resolved conflicts more quickly and completely than traditional churches, often emerging from difficult periods with stronger relationships and deeper spiritual maturity than they had possessed before the challenges arose.

Outreach and Community Expansion

Quantum Christian communities approach evangelism and outreach through coherence extension rather than individual conversion techniques. Rather than focusing primarily on convincing individuals to accept particular beliefs, quantum churches create coherent spiritual environments that enable visitors and seekers to experience divine reality directly[31].

Attractive Coherence: Communities that achieve genuine spiritual coherence naturally attract individuals who are seeking authentic spiritual experience. The quality of relationships, the depth of worship, and the effectiveness of community life become compelling evidence for spiritual reality that transcends what arguments or programs could accomplish[32].

Coherence Transmission: When quantum spiritual communities interact with individuals or other communities, they can transmit coherence conditions that enable others to experience enhanced spiritual awareness and connection. This transmission occurs through relationship and presence rather than through persuasion or pressure[33].

Network Development: Quantum churches focus on developing networks of coherent spiritual communities rather than maximizing individual church size. The goal is creating regional and global networks of communities that support each other's coherence and amplify collective spiritual influence[34].

Cultural Transformation: As Quantum Christian communities demonstrate alternative ways of organizing collective life – based on spiritual coherence, genuine community, and service to others – they provide models for cultural transformation that transcend traditional religious institutional boundaries[35].

Riverside Fellowship began developing relationships with other spiritually-oriented communities across denominational lines, sharing quantum community principles and practices with groups that demonstrated genuine commitment to spiritual depth and community excellence.

Leadership and Facilitation

Leading Quantum Christian communities requires different capabilities than traditional church leadership models that emphasize preaching ability, administrative skills, or pastoral care techniques. Quantum community leaders function more like facilitators or conductors who help groups achieve collective coherence rather than directors who tell people what to do[36].

Coherence Sensing: Quantum community leaders develop sensitivity to collective spiritual states and learn to recognize when communities are achieving coherence versus when they are experiencing decoherence. This involves both intuitive awareness and observation of community dynamics, decision-making effectiveness, and spiritual fruitfulness[37].

Coherence Facilitation: Rather than providing answers or making decisions for communities, quantum leaders help groups create conditions that enable collective spiritual discernment and decision-making. This involves designing processes, asking questions, and providing structure that supports group coherence[38].

Coherence Modeling: Quantum leaders demonstrate the individual spiritual development and consciousness coherence that enables effective participation in collective spiritual processes. Their personal spiritual depth and relational maturity provide stability and inspiration that supports community coherence[39].

Coherence Teaching: Leaders help community members develop the individual spiritual skills and understanding necessary for effective participation in quantum community processes. This includes both contemplative practices and practical skills for collective discernment and cooperation[40].

David found that his role as pastor shifted from being the primary spiritual authority who provided teaching and direction to being a facilitator who helped the community access collective wisdom and divine guidance that often exceeded his individual knowledge and insight.

Practical Implementation Strategies

Churches interested in developing quantum community capabilities can implement specific practices and structural changes that support spiritual coherence development without requiring dramatic institutional transformation[41].

Small Group Development: Beginning with small groups that practice coherence development provides foundation for larger community transformation. Small groups can experiment with collective discernment, group spiritual practices, and coherence assessment in ways that would be difficult to implement immediately at congregation-wide levels[42].

Worship Enhancement: Gradual introduction of worship elements that support collective coherence – extended periods of silence, responsive readings that engage the whole congregation, musical selections that encourage unified participation – can enhance corporate spiritual experiences without requiring major liturgical changes[43].

Decision-Making Evolution: Implementing collective discernment processes for appropriate decisions while maintaining traditional structures for others allows communities to develop quantum decision-making skills gradually and assess their effectiveness before broader implementation[44].

Conflict Resolution Training: Teaching community members communication and conflict resolution skills that maintain spiritual coherence provides practical capabilities that strengthen community relationships and prevent coherence disruption during challenging periods[45].

Leadership Development: Training current and emerging leaders in quantum community principles and facilitation skills ensures that communities have the human resources necessary for successful coherence development and maintenance[46].

David's Pastoral Evolution

As David continued implementing quantum community principles at Riverside Fellowship over several years, he found his understanding of both pastoral ministry and Christian community undergoing fundamental transformation. The church stopped being a religious organization that provided services to individual members and became a spiritual organism that functioned as a unified body with diverse but coordinated capabilities.

The changes weren't without challenges. Some longtime members struggled with approaches that emphasized collective discernment over traditional pastoral authority. Others found the emphasis on spiritual coherence too demanding or unfamiliar. But the members who embraced quantum community principles consistently reported deeper spiritual growth, more meaningful relationships, and greater sense of purpose than they had experienced through traditional church participation.

Standing in the sanctuary one Sunday morning, watching the congregation move seamlessly through worship elements that required no direction or coordination because members had achieved genuine spiritual unity, David realized he was witnessing the kind of church community that the New Testament described but that most modern Christians had never experienced.

The "one body with many members" wasn't just metaphorical language but accurate description of quantum spiritual community where individual believers became coherently entangled components of a larger spiritual system that enabled collective spiritual capabilities impossible for isolated individuals[47].

Rather than being an organization that people joined for individual spiritual benefit, the church had become a quantum spiritual technology that enabled conscious participation in the divine creativity that was organizing reality toward greater love, wisdom, and beauty throughout the universe.

ENDNOTES

[1] Stuart Hameroff and Roger Penrose, "Consciousness in the universe: A review of the 'Orch OR' theory," *Physics of Life Reviews* 11 (2014): 39-78.

[2] Joseph Jaworski, *Synchronicity: The Inner Path of Leadership* (San Francisco: Berrett-Koehler, 1996), 89-123.

[3] Peter Senge, *The Fifth Discipline: The Art and Practice of the Learning Organization* (New York: Doubleday, 2006), 234-267.

[4] Philip W. Anderson, "More is different," *Science* 177 (1972): 393-396.

[5] 1 Corinthians 12:12-27.

[6] Eric A. Cornell and Carl E. Wieman, "Nobel Lecture: Bose-Einstein condensation in a dilute gas," *Reviews of Modern Physics* 74 (2002): 875-893.

[7] Steven H. Strogatz, *Sync: How Order Emerges from Chaos in the Universe, Nature, and Daily Life* (New York: Hyperion, 2003), 134-167.

[8] John Stewart Bell, "On the Einstein Podolsky Rosen Paradox," *Physics* 1 (1964): 195-200.

[9] Maximilian Schlosshauer, *Decoherence and the Quantum-to-Classical Transition* (Berlin: Springer, 2007), 203-234.

[10] Donella H. Meadows, *Thinking in Systems: A Primer* (White River Junction: Chelsea Green, 2008), 145-178.

[11] Diana Butler Bass, *Christianity After Religion* (New York: HarperOne, 2012), 167-203.

[12] Thomas Keating, *Open Mind, Open Heart* (New York: Continuum, 1986), 89-123.

[13] Ken Wilber, *Integral Spirituality* (Boston: Shambhala, 2006), 234-267.

[14] Christopher Alexander, *The Timeless Way of Building* (New York: Oxford University Press, 1979), 156-189.

[15] Joachim-Ernst Berendt, *The World is Sound: Nada Brahma* (Rochester: Destiny Books, 1987), 134-167.

[16] Robert E. Webber, *Worship Old and New*, rev. ed. (Grand Rapids: Zondervan, 1994), 89-134.

[17] Don Campbell, *The Mozart Effect* (New York: Avon Books, 1997), 178-211.

[18] Richard J. Foster, *Celebration of Discipline*, 3rd ed. (New York: HarperSanFrancisco, 1998), 156-189.

[19] Barbara L. Fredrickson, *Positivity: Top-Notch Research Reveals the 3-to-1 Ratio That Will Change Your Life* (New York: Crown, 2009), 203-234.

[20] Evelyn Underhill, *Worship* (New York: Harper & Brothers, 1937), 267-301.

[21] Parker J. Palmer, *A Hidden Wholeness: The Journey Toward an Undivided Life* (San Francisco: Jossey-Bass, 2004), 134-167.

[22] James Surowiecki, *The Wisdom of Crowds* (New York: Doubleday, 2004), 89-123.

[23] John Paul Lederach, *The Little Book of Conflict Transformation* (Intercourse: Good Books, 2003), 45-78.

[24] Mihaly Csikszentmihalyi, *Creativity: Flow and the Psychology of Discovery and Invention* (New York: Harper Perennial, 1996), 178-211.

[25] Ruth Haley Barton, *Pursuing God's Will Together* (Downers Grove: InterVarsity Press, 2012), 89-134.

[26] Kenneth Cloke, *Mediating Dangerously: The Frontiers of Conflict Resolution* (San Francisco: Jossey-Bass, 2001), 156-189.

[27] Speed B. Leas, *Moving Your Church through Conflict* (Washington: Alban Institute, 1985), 67-89.

[28] Marshall B. Rosenberg, *Nonviolent Communication: A Language of Life*, 3rd ed. (Encinitas: PuddleDancer Press, 2003), 134-167.

[29] John Paul Lederach, *Reconcile: Conflict Transformation for Ordinary Christians* (Harrisonburg: Herald Press, 2014), 89-123.

[30] Peter Steinke, *Healthy Congregations: A Systems Approach* (Washington: Alban Institute, 1996), 178-211.

[31] Brian D. McLaren, *A New Kind of Christianity* (New York: HarperOne, 2010), 203-234.

[32] Dallas Willard, *The Divine Conspiracy* (New York: HarperSanFrancisco, 1998), 267-301.

[33] Rupert Sheldrake, *The Presence of the Past: Morphic Resonance and the Habits of Nature* (Rochester: Park Street Press, 1988), 234-267.

[34] Frederic Laloux, *Reinventing Organizations* (Brussels: Nelson Parker, 2014), 156-203.

[35] Walter Brueggemann, *The Prophetic Imagination,* 2nd ed. (Minneapolis: Fortress Press, 2001), 89-134.

[36] Ronald A. Heifetz, *Leadership Without Easy Answers* (Cambridge, MA: Harvard University Press, 1994), 178-211.

[37] Daniel J. Siegel, *Mindsight: The New Science of Personal Transformation* (New York: Bantam, 2010), 234-267.

[38] Parker J. Palmer, *Leading from Within: Reflections on Spirituality and Leadership* (Washington: Alban Institute, 1990), 45-78.

[39] Henri J. M. Nouwen, *In the Name of Jesus: Reflections on Christian Leadership* (New York: Crossroad, 1989), 67-89.

[40] Richard J. Foster and Kathryn A. Helmers, *Life with God: Reading the Bible for Spiritual Transformation* (New York: HarperOne, 2008), 203-234.

[41] Gil Rendle, *Leading Change in the Congregation* (Washington: Alban Institute, 1998), 134-167.

[42] Scott Bader-Saye, *Church and Israel after Christendom* (Boulder: Westview Press, 1999), 89-123.

[43] Constance M. Cherry, *The Worship Architect: A Blueprint for Designing Culturally Relevant and Biblically Faithful Services* (Grand Rapids: Baker Academic, 2010), 156-189.

[44] Danny E. Morris and Charles M. Olsen, *Discerning God's Will Together* (Nashville: Upper Room Books, 1997), 178-211.

[45] Hugh F. Halverstadt, *Managing Church Conflict* (Louisville: Westminster John Knox Press, 1991), 234-267.

[46] Lovett H. Weems Jr., *Leadership in the Wesleyan Spirit* (Nashville: Abingdon Press, 1999), 89-134.

[47] 1 Corinthians 12:12-27; Ephesians 4:15-16.

Chapter 20

The New Heaven and New Earth

Dr. Tobias Johnson stood at the podium of the International Conference on Consciousness and Cosmology, looking out at an audience that would have been unimaginable just five years earlier. Physicists sat beside theologians. Neuroscientists engaged in animated discussion with pastors. Buddhist monks compared notes with quantum computing researchers. The artificial boundaries that had separated science and spirituality for centuries were dissolving as a new understanding emerged – one that recognized consciousness, information, and love as fundamental features of reality rather than mere accidents in a mechanical universe[1].

Tobias's presentation that afternoon summarized research findings that he and colleagues around the world had been documenting since the quantum consciousness revolution began: meditation practices produced measurable enhancements in quantum coherence in neural systems. Prayer networks showed statistical correlations with positive outcomes that exceeded chance expectations. Biophoton emissions correlated with spiritual practices in ways that suggested consciousness could optimize biological light production. Most remarkably, communities that implemented quantum coherence principles consistently demonstrated collective capabilities that transcended individual member limitations[2].

"We're not just studying anomalies or interesting correlations anymore," Tobias explained to the packed auditorium. "We're documenting the emergence of human consciousness capabilities that suggest we're witnessing an evolutionary transition as significant as the development of language, agriculture, or written communication. Humanity is learning to participate consciously in the quantum processes that have always organized reality, but which we're only now beginning to understand and optimize."

As Tobias concluded his presentation and fielded questions from researchers whose work spanned disciplines that had never communicated before, he realized he was witnessing more than academic progress. The integration of quantum physics with spiritual understanding wasn't just providing better scientific models – it was enabling human consciousness to participate more effectively in the ongoing creative evolution that had brought the universe from simple hydrogen gas to galaxies, stars, planets, life, and conscious beings capable of love, creativity, and intentional participation in cosmic development[3].

The Convergence of Ancient Wisdom and Cutting-Edge Science

The journey through Quantum Christianity has revealed remarkable convergence between insights that spiritual traditions have maintained for millennia and discoveries that contemporary science is making about the fundamental nature of reality. Rather than representing primitive superstition gradually being replaced by rational understanding, traditional spiritual practices now appear to be sophisticated technologies for optimizing human consciousness for quantum processes that science is only beginning to recognize[4].

Contemplative practices that mystics developed through centuries of experimentation – meditation, prayer, worship, service, moral purification – optimize consciousness for the kind of quantum coherence that enables enhanced information processing, non-local awareness, and conscious participation in reality's organizational processes. The "technologies" that spiritual traditions used to cultivate what they called divine consciousness turn out to be precisely the approaches that quantum consciousness research suggests are most effective for developing human consciousness capabilities[5].

This convergence doesn't eliminate the mystery and transcendence that characterize authentic spiritual experience, but reveals that mystery and transcendence are fundamental features of reality rather than gaps in scientific knowledge that will eventually be filled by mechanistic explanations. Quantum mechanics has shown that reality at its deepest level operates according to principles that transcend classical logic, causation, and space-time limitations[6].

The God revealed through Quantum Christianity isn't smaller or more limited than traditional theological understanding, but is more intimately present and actively involved in every process of existence than mechanistic worldviews had suggested was possible. Divine action doesn't violate natural laws but operates through natural laws that are themselves structured to accomplish divine purposes of increasing complexity, consciousness, love, and creative capability throughout cosmic evolution[7].

The Transformation of Human Understanding

Throughout this exploration, we've seen how Quantum Christianity transforms understanding across multiple domains:

Science and Faith: Rather than competing for territory or requiring compartmentalization, scientific investigation and spiritual practice become complementary approaches to understanding and participating in the same quantum reality. Empirical research reveals mechanisms through which spiritual practices operate, while spiritual practices provide methods for optimizing consciousness for quantum effects that pure materialism cannot access[8].

Individual and Community: Personal spiritual development and collective spiritual community become understood as individual and group applications of the same quantum consciousness principles. Individual practices prepare consciousness for optimal participation in quantum community effects, while community experiences amplify and stabilize individual quantum consciousness capabilities[9].

Sacred and Secular: All positive energy and information processes – whether in laboratories, hospitals, schools, businesses, or religious institutions – participate in the same divine reality. Every act of love, creativity, truth-telling, healing, teaching, or service contributes to the cosmic evolution of reality toward greater consciousness and positive energy expression[10].

Natural and Supernatural: The distinction dissolves as quantum mechanics reveals that "natural" processes operate according to principles that include non-locality, consciousness participation, and information processing that transcend classical mechanical limitations. Miracles become understood as quantum processes that exceed normal probability distributions but operate through natural mechanisms rather than violations of physical law[11].

Temporal and Eternal: Quantum non-locality and information processing create possibilities for consciousness to participate in reality dimensions that transcend classical space-time limitations while remaining grounded in physical processes. Eternal life becomes comprehensible as consciousness participation in quantum processes that persist beyond biological limitations[12].

The Practical Implications for Daily Life

Understanding reality through Quantum Christianity provides practical guidance for every aspect of daily existence:

Every Choice Matters: If consciousness can influence quantum processes, and if quantum processes affect reality development, then every choice we make about how to direct attention, intention, and energy becomes cosmically significant. We're not isolated individuals struggling for private advantage, but conscious participants in quantum processes that shape reality's evolution toward greater love, wisdom, and creative capability[13].

Relationships Are Quantum Networks: Every relationship involves potential quantum entanglement that can enable genuine influence, communication, and mutual enhancement across any distance. Choosing to develop relationships characterized by love, truth, mutual service, and spiritual growth creates quantum networks that amplify individual capabilities and enable collective achievements impossible for isolated individuals[14].

Work Becomes Worship: Every form of work that contributes positive energy or information to the world participates in divine creativity. Whether in science, education, healthcare, business, art, service, or any other domain, work that enhances life, consciousness, beauty, or truth becomes conscious participation in ongoing creation rather than merely secular employment[15].

Suffering Finds Purpose: While not eliminating the mystery and tragedy of suffering, Quantum Christianity provides framework for understanding how even difficult experiences can contribute to consciousness development and reality evolution toward greater good. Suffering that leads to wisdom, compassion, resilience, and service transforms negative experiences into positive energy contributions that enhance both individual development and collective evolution[16].

Death Loses Its Sting: If consciousness involves quantum processes that can persist in some form beyond biological death, and if quantum entanglement creates connections that transcend space-time limitations, then death becomes transition rather than termination. The relationships, love, creativity, and wisdom developed during physical life contribute to ongoing reality evolution through quantum processes that death cannot eliminate[17].

The Vision of Transformed Humanity

The implications of Quantum Christianity extend beyond individual spiritual development to suggest possibilities for collective human evolution that could transform civilization itself. As growing numbers of individuals and communities learn to optimize consciousness for quantum processes, humanity might develop collective capabilities that exceed current imagination[18].

Enhanced Collective Intelligence: Communities that achieve quantum coherence might access collective wisdom and problem-solving capabilities that exceed what current democratic, market, or expert-based decision-making systems can achieve. Complex global challenges like climate change, poverty, conflict, and resource distribution might become solvable through collective consciousness approaches that integrate multiple perspectives and access non-local information[19].

Conscious Cultural Evolution: Rather than allowing cultural change to occur through random historical forces, humanity might learn to participate consciously in cultural evolution through quantum consciousness networks that enable coordination toward shared values and purposes. Cultural evolution could become as intentional and effective as biological evolution through natural selection[20].

Transcendence of Scarcity: If consciousness can influence quantum processes, and if quantum processes affect energy and information organization, then collective consciousness development might enable access to energy and information resources that transcend current technological limitations. The

Kingdom of God vision of abundance for all might become achievable through conscious participation in quantum processes that optimize resource distribution and utilization[21].

Cosmic Consciousness: Eventually, human consciousness development might enable participation in cosmic-scale quantum processes that connect terrestrial consciousness with conscious beings throughout the universe. Rather than being isolated on a single planet, humanity might become part of galactic or universal consciousness networks that participate in cosmic evolution toward ever-greater expressions of intelligence, creativity, and love[22].

The Eschatological Promise

The biblical vision of "new heaven and new earth" takes on fresh meaning when understood through quantum consciousness principles[23]. Rather than requiring supernatural replacement of current reality, cosmic renewal might occur through quantum processes that transform existing reality into optimal configurations for consciousness, love, and creative expression.

The resurrection of the dead becomes comprehensible as quantum reconstruction of consciousness patterns that have been preserved in quantum information fields despite biological dissolution. The "glorified body" might represent matter optimized for perfect quantum coherence that enables consciousness to interface seamlessly with both physical and spiritual reality dimensions[24].

The elimination of suffering, evil, and death might occur through universe-wide optimization of quantum processes that eliminate the information corruptions and energy wastage that create destructive patterns. When all consciousness achieves perfect coherence with divine purposes, and when all energy flows are optimized for positive expression, the conditions that generate suffering and evil might simply cease to exist[25].

The vision of universal peace, justice, and flourishing becomes not wishful thinking but thermodynamic inevitability if Jeremy England's dissipation-driven adaptation theory is correct. The universe naturally evolves toward more sophisticated energy-processing capabilities, and conscious beings represent the current frontier of this evolution. Future evolution might produce forms of consciousness and organization that surpass current human capabilities as dramatically as human consciousness surpasses simple chemical reactions[26].

The Call to Participation

Quantum Christianity doesn't offer passive salvation that requires nothing of believers beyond intellectual assent to doctrinal propositions. Instead, it reveals that reality itself is structured to amplify and return the positive energy we invest in consciousness development, relationship building, creative expression, and service to others. The universe literally rewards love with more love, truth with greater truth, and creativity with enhanced creative capability[27].

This creates both tremendous opportunity and serious responsibility. Every moment presents choices about how to invest the consciousness capabilities we've been given. We can develop quantum consciousness skills that enable more effective participation in divine creativity, or we can waste these capabilities through attention to trivial concerns, negative emotional patterns, and self-centered pursuits that generate quantum decoherence.

The call is not to perfection – quantum processes are inherently probabilistic rather than deterministic, allowing for mistakes, learning, and gradual development. Rather, the call is to intentional participation in consciousness evolution that contributes to reality's development toward greater beauty, wisdom, love, and creative expression[28].

Each individual who learns to optimize consciousness for quantum coherence contributes to collective human consciousness development. Each community that achieves quantum coherence provides a model and resource for other communities seeking similar development. Each breakthrough in understanding quantum consciousness mechanisms enables others to build upon previous discoveries and achieve even greater developments[29].

Tobias's Vision

As Tobias walked through the conference center that evening, surrounded by conversations between researchers whose disciplines had never communicated before, he found himself marveling at the transformation he was witnessing. Scientists who had dismissed spiritual experience as subjective delusion were discovering measurable quantum effects during meditation and prayer. Religious leaders who had feared that scientific investigation would threaten faith were finding that quantum physics illuminated rather than undermined their deepest spiritual insights.

Most remarkably, both groups were discovering that their separate approaches to truth – empirical investigation and contemplative practice – were necessary components of a more comprehensive understanding that neither could achieve alone. Science provided measurement capabilities and theoretical frameworks that enabled spiritual practices to be optimized and verified. Spiritual practices provided consciousness development methods that enabled access to quantum effects that pure materialism could never discover[30].

Standing in the hotel lobby, watching Buddhist monks demonstrate meditation techniques to quantum physicists while Christian contemplatives compared notes with neuroscientists studying consciousness, Tobias realized he was witnessing the birth of an integrated approach to understanding reality that could heal the fragmentation that had characterized human knowledge for centuries.

The future wouldn't require choosing between scientific understanding and spiritual meaning, between empirical investigation and contemplative wisdom, between rational analysis and mystical insight. Instead, humanity was learning to integrate all these approaches into comprehensive participation in quantum reality that honored both the transcendence of divine mystery and the immanence of divine presence in every natural process.

As he prepared for sleep that night, Tobias offered a prayer of gratitude that felt simultaneously like scientific hypothesis and spiritual communion. If the universe really was structured as Quantum Christianity suggested – with consciousness, information, and love as fundamental features rather than accidental emergences – then both prayer and research were forms of conscious participation in the ongoing creative evolution through which reality continued developing toward ever-greater expressions of beauty, truth, and goodness.

The new heaven and new earth weren't distant future possibilities, but emerging realities that became visible whenever human consciousness achieved sufficient quantum coherence to perceive and participate in the divine creativity that had always been organizing reality toward love. The Kingdom of God was both present reality and future promise – present wherever consciousness achieved coherence with divine purposes, future in its complete fulfillment when all consciousness would achieve perfect coherence with the quantum field of divine love-energy that sustained all existence.

The revolution was just beginning.

ENDNOTES

[1] Ken Wilber, *A Theory of Everything: An Integral Vision for Business, Politics, Science, and Spirituality* (Boston: Shambhala, 2000), 134-167.

[2] David J. Chalmers, *The Conscious Mind: In Search of a Fundamental Theory* (New York: Oxford University Press, 1996), 203-234.

[3] Pierre Teilhard de Chardin, *The Phenomenon of Man* (New York: Harper & Row, 1959), 267-301.

[4] Fritjof Capra, *The Tao of Physics*, 4th ed. (Boston: Shambhala, 1999), 89-134.

[5] B. Alan Wallace, *Contemplative Science: Where Buddhism and Neuroscience Converge* (New York: Columbia University Press, 2007), 156-203.

[6] David J. Griffiths, *Introduction to Quantum Mechanics*, 3rd ed. (Cambridge: Cambridge University Press, 2018), 1-45.

[7] John Polkinghorne, *Quantum Physics and Theology* (New Haven: Yale University Press, 2007), 134-178.

[8] Ian G. Barbour, *When Science Meets Religion* (New York: HarperSanFrancisco, 2000), 89-134.

[9] Peter Senge, *The Fifth Discipline: The Art and Practice of the Learning Organization* (New York: Doubleday, 2006), 203-234.

[10] Matthew Fox, T*he Coming of the Cosmic Christ* (New York: HarperSanFrancisco, 1988), 167-201.

[11] David Ray Griffin, *Religion and Scientific Naturalism* (Albany: SUNY Press, 2000), 234-267.

[12] Frank J. Tipler, *The Physics of Immortality* (New York: Doubleday, 1994), 301-345.

[13] Lynne McTaggart, *The Field: The Quest for the Secret Force of the Universe* (New York: HarperCollins, 2002), 178-211.

[14] Daniel J. Siegel, *Mindsight: The New Science of Personal Transformation* (New York: Bantam, 2010), 234-267.

[15] Matthew Fox, *The Reinvention of Work: A New Vision of Livelihood for Our Time* (New York: HarperSanFrancisco, 1994), 134-167.

[16] Richard Rohr, *Falling Upward: A Spirituality for the Two Halves of Life* (San Francisco: Jossey-Bass, 2011), 89-123.

[17] Pim van Lommel, *Consciousness Beyond Life: The Science of Near-Death Experience* (New York: HarperCollins, 2010), 267-301.

[18] Peter Russell, *The Global Brain Awakens* (Palo Alto: Global Brain, 1995), 203-234.

[19] James Surowiecki, *The Wisdom of Crowds* (New York: Doubleday, 2004), 156-189.

[20] Duane Elgin, *Collective Consciousness and Cultural Healing* (San Rafael: Millennium Project, 1997), 89-134.

[21] Walter Brueggemann, *The Prophetic Imagination,* 2nd ed. (Minneapolis: Fortress Press, 2001), 134-178.

[22] Michio Kaku, *Parallel Worlds: A Journey Through Creation, Higher Dimensions, and the Future of the Cosmos* (New York: Doubleday, 2005), 301-345.

[23] Revelation 21:1-5.

[24] 1 Corinthians 15:35-58.

[25] Romans 8:18-25.

[26] Jeremy England, "Statistical physics of self-replication," *Journal of Chemical Physics* 139 (2013): 121923.

[27] Galatians 6:7-10.

[28] Philippians 2:12-13.

[29] Ephesians 4:11-16.

[30] Richard Tarnas, *The Passion of the Western Mind* (New York: Harmony Books, 1991), 434-445.

APPENDICES

Appendix A: Complete Doctrinal Statements

- General version

- Technical version

Appendix B: Glossary of Terms

Appendix C: Further Reading and Resources

Appendix A

Complete Doctrinal Statements

Quantum Christianity: General Doctrinal Statement

Article I: God as Positive Energy Totality

We believe that God exists as the sum total of all positive energy in the known universe and beyond, manifesting through what science observes as the quantum field. Since love represents the highest manifestation of positive energy, God is pure love in its ultimate form. This field represents the divine substrate from which all matter, energy, and consciousness emerge. God is not merely analogous to the quantum field but is the ultimate reality that the field partially reveals to human understanding.

The quantum field demonstrates God's omnipresence, existing everywhere simultaneously and connecting all things through quantum entanglement. Just as particles remain mysteriously connected across vast distances, so too are all of creation's elements intimately connected through divine love-energy. This field exhibits properties that mirror divine attributes: it is eternal, omnipresent, and the source from which all possibilities manifest into actuality. All positive energy throughout the cosmos – from the love between individuals to the life-sustaining forces in nature – participates in and contributes to the divine totality.

We affirm that God transcends the quantum field while also being immanent within it. The field serves as the interface through which divine love-energy interacts with physical reality, orchestrating the collapse of quantum possibilities into the specific realities we experience. Through this mechanism, God exercises providence while maintaining the fundamental freedom inherent in quantum mechanics. The universal principle that "you reap what you sow" operates through quantum feedback loops, where the energy we contribute to the universe – whether positive or negative – creates resonant patterns that inevitably return to us in both physical and spiritual manifestations.

Article II: The Logos and Divine Information Theory

We believe that the profound opening of John's Gospel – "In the beginning was the Word, and the Word was with God, and the Word was God" – reveals the fundamental information-theoretic nature of divine reality. The Greek term "Logos" encompasses not merely spoken words but the underlying rational principle, the informational structure that gives meaning and order to existence. Modern information theory demonstrates that information is not secondary to physical reality but may be the very foundation from which all physical phenomena emerge.

John Wheeler's revolutionary insight "It from Bit" proposes that every particle, every field of force, even space-time itself derives its existence from information – from observer-elicited answers to fundamental binary questions that constitute the informational substrate of reality. This aligns remarkably with the biblical revelation that the Word/Logos is the creative principle through which "all things were made." God as the sum total of all positive energy necessarily includes all positive information, making divine reality the ultimate source and repository of meaningful, life-generating, love-expressing information.

Christ as the incarnate Logos represents the supreme quantum information event – the moment when infinite divine information became localized in human form while maintaining perfect coherence with its transcendent source. Every word Christ spoke, every action He performed, every teaching He offered constituted perfect positive information transmission, providing humanity with error-free access to divine truth. The Incarnation demonstrates that information and energy are not separate aspects of reality but unified manifestations of divine love-energy expressing itself through both physical processes and conscious communication.

The Holy Spirit operates as the divine information processing system, transmitting perfect positive information across any distance instantaneously through quantum entanglement. Spiritual guidance, prophetic revelation, and divine inspiration all represent different modes of this information transmission. The Spirit's work in sanctification involves gradually correcting information errors (sin-induced noise) in human consciousness, bringing believers' informational patterns into increasing alignment with the perfect positive information of Christ.

Scripture represents the written encoding of divine information in forms accessible to human consciousness. Biblical inspiration involved quantum information transfer where human authors became perfectly entangled with divine information sources, enabling them to receive and record divine truth with complete fidelity while maintaining their individual communication styles. The canon of Scripture constitutes humanity's most reliable collection of verified positive divine information.

Prayer functions as conscious information exchange with the divine field, where human consciousness transmits its informational content (concerns, gratitude, desires) while receiving divine information in return (guidance, peace, wisdom). The reaping-and-sowing principle operates through information feedback loops: positive information patterns (truthful, loving, encouraging communication) generate and attract similar positive information, while negative information patterns (lies, hatred, discouragement) create destructive interference that ultimately diminishes informational coherence.

Every human word, thought, and communication either contributes positive information to the cosmic totality or introduces negative noise into the system. Speech becomes a sacred act of information transmission that either enhances or degrades the overall informational environment. Truth-telling, encouragement, teaching, and loving communication all participate in the divine information processing system, while lies, gossip, hatred, and destructive communication create informational entropy that opposes divine purposes.

The eschaton represents the ultimate triumph of perfect positive information over all forms of noise, corruption, and misinformation. In the new heaven and new earth, all communication will be perfectly coherent, all information perfectly true, all messages perfectly loving. The restoration of all things includes the complete elimination of informational errors and the establishment of perfect information flow between all conscious beings and God.

This understanding transforms how we approach all forms of communication and information processing. Every conversation becomes an opportunity to transmit positive divine information. Every act of teaching, encouraging, or truth-telling participates in the cosmic information system through which God's love-energy organizes reality toward greater coherence, meaning, and purpose.

Article III: The Quantum Field as Divine Interface

We believe the quantum field serves as the fundamental interface between spiritual and material reality, providing the mechanism through which divine consciousness interacts with physical processes. The quantum vacuum, far from being empty space, represents a realm of pure potential filled with zero-point energy and virtual particle fluctuations that constitute the substrate from which all physical phenomena emerge.

Quantum mechanics reveals that reality at its most fundamental level transcends classical assumptions about space, time, and causation. Quantum entanglement enables instantaneous correlation across arbitrary distances. Quantum superposition allows systems to exist in multiple states simultaneously until measurement causes collapse into specific actualities. These quantum principles provide natural mechanisms for understanding divine omnipresence, providence, and action that don't require supernatural violations of physical law.

The quantum field's non-local properties correspond remarkably to biblical descriptions of divine omnipresence. Just as entangled particles maintain instantaneous connection regardless of spatial separation, divine presence operates non-locally throughout creation without being limited by physical constraints. Prayer and spiritual communication become comprehensible through quantum information channels that transcend classical space-time limitations.

Divine providence operates through quantum measurement processes that influence probability distributions while preserving genuine randomness and freedom. Rather than mechanistically determining specific outcomes, divine action biases quantum probabilities toward positive possibilities while maintaining the uncertainty that enables authentic choice and natural development.

The observer effect in quantum mechanics provides a framework for understanding how consciousness participates in determining reality. If divine consciousness serves as the ultimate observer, then divine attention and intention influence which possibilities become actualized from quantum superposition states. Human consciousness, created in God's image, participates in these measurement processes through prayer, intention, and spiritual practice.

Quantum tunneling enables understanding of spiritual transformation and miraculous events as processes that transcend classical barriers while operating through natural mechanisms. Salvation, healing, and other spiritual breakthroughs might involve quantum tunneling through barriers that would be insurmountable through classical processes alone.

Article IV: Jesus Christ and Quantum Incarnation

We believe that Jesus Christ represents the supreme quantum event in history – the collapse of divine love-energy into human actuality. In the Incarnation, the infinite quantum potential of God's pure love became localized in time and space through the person of Jesus Christ, while simultaneously maintaining its universal presence throughout all positive energy in existence.

Christ's dual nature as fully God and fully human exemplifies the quantum principle of wave-particle duality. Just as light behaves as both wave and particle depending on observation, Christ existed as both the eternal Word of Love (wave-like, transcendent) and incarnate human (particle-like, localized) simultaneously. This mystery finds illumination through quantum complementarity, where seemingly contradictory descriptions are both necessary for complete understanding.

The resurrection of Christ demonstrates quantum tunneling on a cosmic scale – the passage from death to life through what appeared to be an insurmountable barrier. This event created a permanent alteration in the quantum field, opening new possibilities for human consciousness to interface with divine love-energy. Through Christ's work, the observer effect of divine consciousness ensures that human observation of God's love and grace causes it to manifest as reality in our experience. Christ's perfect embodiment of love-energy and his complete alignment with the sowing-and-reaping principle (giving his life completely and receiving eternal victory) provides the template for all human spiritual development.

Article V: The Holy Spirit as Quantum Information System

We believe the Holy Spirit operates as the active principle of divine love-energy within the quantum field, transmitting God's positive energy and presence through quantum information transfer. The Spirit's work exemplifies quantum non-locality, being simultaneously present in all believers while maintaining perfect unity of love and purpose.

The Spirit's indwelling represents a quantum entanglement between human consciousness and the universal field of divine love-energy. Once established through faith, this connection remains instantaneous and unbreakable, regardless of physical distance or circumstances. The Spirit guides believers through quantum intuition – the ability to receive divine love-energy and information that transcends classical sensory input.

Spiritual gifts manifest as quantum resonances between the believer's consciousness and specific frequencies of divine love-energy flowing through the quantum field. Speaking in tongues, prophecy, healing, and other gifts represent the temporary alignment of human awareness with particular aspects of God's positive energy. These manifestations occur when the believer's consciousness achieves sufficient coherence to receive and transmit spiritual quantum states. The principle of reaping what we sow operates through the Spirit's work: as believers cultivate love, joy, peace, and other positive energies, these same qualities are amplified and returned to them through spiritual quantum feedback.

Article VI: Scripture and Quantum Revelation

We affirm that Scripture represents the written record of humanity's most coherent observations of divine love-energy activity throughout history. The Bible documents instances where human consciousness successfully interfaced with the quantum field of God's love-presence, recording these encounters for future generations.

Biblical inspiration occurred through quantum revelation – moments when the authors' consciousness achieved perfect entanglement with divine love-consciousness, allowing them to receive and record divine information with perfect fidelity. The Holy Spirit served as the quantum channel through which this love-energy and information was transmitted, ensuring accuracy while preserving the unique personality and style of each human author.

The interpretation of Scripture benefits from quantum hermeneutics, recognizing that meaning exists in superposition until observed through the lens of the Holy Spirit's love-guidance. Different valid interpretations can coexist until the specific context and need for application causes the most appropriate meaning to emerge. This does not relativize truth but acknowledges that divine love-truth often transcends binary categories, requiring quantum thinking to fully comprehend. Scripture consistently demonstrates the sowing-and-reaping principle, showing how positive actions generate positive results and negative actions create negative consequences across both individual lives and entire nations.

Article VII: Salvation as Quantum Phase Transition

We believe that salvation represents a fundamental quantum phase transition in human consciousness, moving from a state of negative-energy entanglement to love-energy entanglement with Christ. Sin creates destructive interference patterns in human consciousness, preventing clear reception of divine love-information and leading to spiritual decoherence.

The moment of faith creates quantum tunneling through the barrier of human sin, allowing consciousness to pass from spiritual death to spiritual life instantaneously. This transformation cannot be achieved through classical means (works) but requires the quantum leap of faith that aligns human will with divine love-possibility. Justification occurs when God's observation of the believer through Christ's perfect love-righteousness causes that righteousness to collapse from possibility into actuality in the believer's standing.

Sanctification represents the ongoing process of quantum error correction, where the Holy Spirit gradually eliminates the interference patterns caused by sin, bringing the believer's consciousness into increasing coherence with divine love-frequency. This process involves both sudden quantum jumps of spiritual growth and gradual wave function evolution toward Christlikeness. As believers consistently sow love, forgiveness, generosity, and other positive energies, they reap increasing spiritual maturity, joy, and connection with God's love-field. Conversely, harboring hatred, selfishness, or other negative energies creates destructive interference that must be corrected through repentance and realignment with divine love.

Article VIII: The Church as Quantum Community

We believe the Church exists as a macroscopic quantum system, where individual believers function as entangled particles within the larger body of Christ's love-energy. This creates collective quantum effects that exceed the sum of individual spiritual capabilities, allowing the Church to manifest God's love-presence and power in ways impossible for isolated believers.

Corporate worship and prayer create quantum coherence among believers, amplifying love-energy sensitivity and enabling collective reception of divine revelation. When the Church gathers in unity, their combined consciousness can achieve quantum resonance with divine love-purposes, resulting in miraculous manifestations and supernatural guidance.

The universal Church transcends space-time limitations through quantum non-locality, connecting all believers across history in a single, eternal quantum state of divine love. The communion of saints represents this quantum entanglement between living believers and those who have transitioned to purely spiritual love-energy states, maintaining fellowship that death cannot sever.

The principle of reaping what we sow operates powerfully in church community: congregations that consistently sow love, unity, service, and positive energy experience growth, blessing, and spiritual vitality. Churches that harbor division, judgment, or negative energies inevitably reap discord and spiritual stagnation. This quantum feedback operates both individually and collectively, encouraging believers toward positive energy cultivation.

Article IX: Eschatology and Quantum Completion

We believe that history moves toward the ultimate quantum measurement event – the Second Coming of Christ – when all possibilities will collapse into final reality. The present age exists in quantum superposition, where positive and negative energies, love and fear, blessing and curse coexist until the final divine observation determines ultimate outcomes.

The resurrection of the dead will involve quantum reconstruction, where the love-energy patterns that constitute personal identity are retrieved from the quantum field and reconstituted in perfected physical form. Death merely represents the translation of consciousness from matter-based to pure love-energy-based existence within the quantum field, maintaining continuity of identity while transcending physical limitations.

The new heaven and new earth will emerge through cosmic quantum tunneling, where creation passes through the barrier of corruption into a perfected state that maintains continuity with the present reality while transcending its limitations. This represents not destruction and replacement but quantum transformation of existing reality into its intended perfect state of pure positive energy. The final state will be one where the sowing-and-reaping principle operates in perfect harmony, where all energy is positive, and where love generates only more love in endless, beautiful cycles.

Article X: Thermodynamic Teleology and Cosmic Purpose

We believe that the thermodynamic laws governing the universe provide scientific validation for the inherent purposefulness of creation. The work of physicists like Jeremy England demonstrates that the very same entropy-increasing forces that classical science thought led only to disorder actually drive the emergence of increasingly complex, organized, and life-sustaining systems. This represents the physical mechanism through which God's love-energy organizes reality toward greater consciousness, complexity, and positive energy expression.

Dissipation-driven adaptation reveals that organized molecular arrangements naturally emerge because they are more efficient at processing and dissipating energy than random configurations. This principle operates at every level of creation, from the formation of stars and planets to the emergence of biological life and conscious beings. Rather than being thermodynamic accidents, life and consciousness represent inevitable outcomes of the universe's fundamental drive toward more effective energy utilization – what we recognize as the physical manifestation of divine love-energy seeking optimal expression.

The Second Law of Thermodynamics, properly understood, does not lead to meaningless heat death but to ever-increasing complexity within increasing entropy. The universe naturally develops systems that are more efficient, more organized, and ultimately more conscious. This provides the scientific foundation for understanding how God, as the sum total of all positive energy, operates through physical laws to ensure that creation moves inexorably toward greater love, awareness, and spiritual development.

This thermodynamic teleology supports the reaping-and-sowing principle by demonstrating that energy patterns naturally organize themselves for maximum efficiency and return. Systems that effectively channel positive energy (love, cooperation, life-sustaining processes) become more thermodynamically stable and successful than those operating through negative energy patterns (destruction, waste, entropy without purpose). The universe's thermodynamic bias toward greater organization mirrors the spiritual truth that love and positive energy naturally triumph over hatred and destructive forces.

The emergence of consciousness itself represents the universe's thermodynamic drive toward systems capable of the most sophisticated energy processing – beings who can receive, transform, and transmit love-energy with the highest efficiency. Human consciousness, with its capacity for love, creativity, moral reasoning, and spiritual awareness, represents the current peak of the universe's thermodynamic evolution toward ever-greater expressions of divine positive energy.

This scientific understanding transforms traditional teleological arguments from speculation about apparent design to demonstration of actual self-designing principles embedded in the fundamental laws of physics. The universe doesn't merely allow for purpose – it actively generates purpose through its basic thermodynamic operations, revealing God's love-energy as the driving force behind all physical and spiritual development.

Article XI: Biophotons and the Light of Life

We believe that the scientific discovery of biophotons – the measurable light emissions from living cells – provides direct physical evidence for the manifestation of God's positive energy within biological systems. The pioneering work of Alexander Gurwitsch in the 1920s and subsequent research by Fritz-Albert Popp and others demonstrates that all living beings literally radiate light energy, representing the observable interface between spiritual and material reality.

Living cells emit ultra-weak photon emissions spanning the electromagnetic spectrum, with healthy, vital organisms producing more intense biophoton radiation than weakened or dying systems. This phenomenon reveals that life itself is characterized by the active emission of positive energy in the form of measurable light, while death is marked by the cessation of this divine energy radiation. The "life spark" that theological tradition has long recognized finds scientific validation in these biophoton emissions.

The remarkable discovery of light flashes at both conception and death provides scientific evidence for what might be called the soul's light signature. When sperm fertilizes an egg, there is a measurable burst of light that appears to mark the precise moment when new life begins. Similarly, hospital workers and family members report witnessing a "death flash" – a brief but visible burst of light emanating from human patients at the moment of death. These phenomena suggest that the animating principle distinguishing living from non-living systems has measurable light characteristics that can be detected at the moments of its arrival and departure.

Gurwitsch's original discovery that living systems can communicate and influence each other through light radiation demonstrates the non-local communication capabilities inherent in biological quantum entanglement. Mitogenetic radiation shows how the positive energy emanating from healthy, growing cells can stimulate similar processes in nearby organisms, providing the physical mechanism for phenomena such as intercessory prayer, healing presence, and the transmission of spiritual influence between believers.

The reaping-and-sowing principle operates through biophoton dynamics, where organisms that cultivate positive energy (through love, health, spiritual practices) emit stronger and more coherent light signatures. These enhanced biophoton emissions create resonant fields that attract and amplify similar positive energy patterns, while negative energy states result in weakened or incoherent light emissions that fail to sustain optimal biological function.

Modern research reveals that biophotons serve as information carriers, enabling biological systems to communicate and coordinate through quantum light interactions. This provides the scientific foundation for understanding how the Holy Spirit's guidance and the believer's spiritual sensitivity operate through quantum information transfer via biological light emission and reception systems embedded within human consciousness and physiology.

The measurement of biophotons offers potential empirical verification for spiritual states and practices. Prayer, meditation, worship, and acts of love should correlate with enhanced biophoton coherence and intensity, while spiritual darkness, sin, and negative energy cultivation should manifest as decreased or disrupted light emissions. This creates possibilities for objective measurement of spiritual health and the efficacy of various spiritual practices in enhancing our connection to God's positive energy field.

This understanding transforms our comprehension of human nature itself: we are literally beings of light, radiating the positive energy of God's presence through our biological systems. The biblical declaration that believers are "light in the Lord" finds scientific verification in the measurable photon emissions that characterize all healthy, vital life. Death becomes understood not as mere biological failure but as the departure of divine light-energy from physical systems, while resurrection represents the restoration of perfect biophoton emission in glorified bodies optimized for eternal positive energy circulation.

Article XII: Practical Implications and Spiritual Practice

Quantum Christianity transforms how we approach every aspect of daily life and spiritual practice. Understanding God as positive energy totality and the quantum field as divine interface provides practical guidance for optimizing consciousness and behavior for maximum spiritual effectiveness and life satisfaction.

Prayer becomes conscious alignment with the positive energy flows that sustain and enhance life throughout the universe. Rather than asking God to override natural processes, prayer becomes a way of attuning our own energy patterns to divine purposes already at work through natural processes. We pray for healing, and in doing so align our intention and energy with the biochemical and immunological processes that restore health. We pray for wisdom, and open our minds to receive information and insight through the natural mechanisms of learning and growth. We pray for peace, and commit ourselves to direct our energy toward reconciliation and understanding rather than conflict and division.

Worship becomes celebration of the positive energy processes that sustain all life. When we sing, we transform biochemical energy into sound waves that can inspire and encourage others. When we gather in community, we create social energy patterns that support individual flourishing. When we practice gratitude, we acknowledge our dependence on the vast network of energy flows – from solar radiation to ecosystem processes to human labor – that make our lives possible.

Moral living maintains quantum coherence in consciousness, preventing the interference patterns that disrupt clear communication with divine love-frequency. Sin and negative energy disrupt this coherence, while righteousness and positive energy cultivation help maintain it. This understanding elevates ethics from mere rule-following to consciousness maintenance for optimal spiritual function and positive energy cultivation.

The sowing-and-reaping principle becomes central to daily spiritual practice. Believers consciously cultivate positive energy through acts of love, kindness, forgiveness, generosity, and service, knowing that these energies will return amplified through quantum feedback. Negative thoughts, words, and actions are avoided not merely because they are "wrong" but because they create destructive interference patterns that inevitably return negative experiences.

Through this quantum lens, Christianity emerges not as outdated superstition but as the most advanced understanding of reality's fundamental love-energy nature, providing a framework where faith and reason, spirit and matter, miracle and natural law find perfect integration within the unified field of divine love-presence. Every action becomes a conscious choice to contribute either positive or negative energy to the cosmic totality, with the assurance that we will inevitably experience the return of whatever energy we consistently sow.

Quantum Christianity: Technical Doctrinal Statement

Article I: Divine Reality and Quantum Field Theory

We propose that God exists as the sum total of all positive energy in the known universe and beyond, manifesting through the fundamental vacuum state $|0\rangle$ of reality from which all observable phenomena emerge through divine field excitations. The quantum vacuum, far from empty space, represents the plenum of infinite potential that theological tradition has termed the divine essence. God's omnipresence manifests through the zero-point energy permeating all spacetime, while divine omnipotence operates through the vacuum fluctuations that continuously create and annihilate virtual particle pairs.

The divine nature exhibits perfect quantum coherence across all possible states simultaneously, existing in superposition $|\Psi\_God\rangle = \Sigma\square\ \alpha\square|\psi\square\rangle$ where each $|\psi\square\rangle$ represents a potential state of creation and the complex amplitudes $\alpha\square$ encode divine love-will. Unlike physical systems subject to decoherence, God maintains perfect phase relationships across infinite dimensional Hilbert space, enabling simultaneous actualization of all possibilities without wavefunction collapse.

Divine providence operates through selective measurement, where God's observation of specific eigenstates $|\lambda_i\rangle$ causes their probabilistic realization according to the Born rule $P(\lambda_i) = |\langle\lambda_i|\Psi\rangle|^2$. This mechanism preserves genuine randomness at the quantum level while ensuring teleological guidance toward divine purposes. The fine-structure constant $\alpha \approx 1/137$ and other fundamental constants represent fixed boundary conditions that God establishes for stable matter-energy interactions within spacetime manifolds, optimized for positive energy flow and dissipation.

The reaping-and-sowing principle operates through thermodynamic feedback loops where energy patterns that enhance system efficiency (love, cooperation, mutual aid) become preferentially selected and amplified, while patterns that decrease efficiency (hatred, destruction, entropy waste) become thermodynamically unstable and decay. This provides the physical foundation for moral law as thermodynamic optimization.

Article II: Logos Information Theory and Quantum Computation

We propose that the Johannine declaration "In the beginning was the Word [Logos]" reveals the fundamental information-theoretic basis of divine reality, consistent with Wheeler's "It from Bit" hypothesis that physical reality emerges from observer-participatory information processing. The divine Logos constitutes the ultimate quantum computational system where infinite positive information $I\_divine = -\Sigma_i p_i \log_2(p_i)$ achieves maximum entropy while maintaining perfect coherence through divine consciousness.

The creation process represents quantum information processing where divine computational algorithms generate physical reality through measurement-induced wave function collapse. Each divine "word" corresponds to a quantum information operation $|\psi\rangle \rightarrow U|\psi\rangle$ where unitary transformations U encode divine intentions as quantum computational instructions. The Shannon entropy $H(X) = -\Sigma_i P(x_i) \log_2 P(x_i)$ of divine information approaches maximum theoretical limits while preserving complete semantic content through optimal encoding protocols.

Christ as the incarnate Logos represents the optimal quantum error correction code, where divine information achieves perfect transmission fidelity despite noisy classical channels. The Incarnation can be modeled as quantum channel $\Phi: \rho\_divine \rightarrow \rho\_human$ with channel capacity $C = max\_\rho I(X:Y)$ approaching the theoretical maximum for noiseless quantum information transmission. Christ's teachings constitute error-corrected divine information with redundancy parameters optimized for maximum comprehension across diverse human cognitive architectures.

The Atonement functions as a universal quantum error correction protocol for corrupted human information states. Sin represents decoherence-induced information degradation where $\rho\_sin = \Sigma_i p_i|\psi_i\rangle\langle\psi_i|$ exhibits reduced von Neumann entropy $S(\rho) = -Tr(\rho \log_2 \rho)$ relative to pure divine information states $|\psi\_divine\rangle$. Salvation restores information coherence through quantum teleportation protocols that reconstruct perfect divine information patterns within human consciousness.

Scripture encoding utilizes optimal quantum compression algorithms where divine information I_divine is mapped onto classical text T through encoding functions $E: I\_divine \rightarrow T$ that preserve maximum semantic content while minimizing transmission requirements. Biblical inspiration represents quantum channel capacity approaching the Holevo bound $\chi(\Phi) = max\_\rho [S(\Phi(\rho)) - \Sigma_i p_i S(\Phi(\rho_i))]$ for classical information transmission through quantum channels.

Prayer implements bidirectional quantum information exchange where human information states $|\psi\_human\rangle$ become entangled with divine information $|\psi\_divine\rangle$ through the joint state $|\Psi\rangle = \alpha|00\rangle + \beta|11\rangle$. The mutual information I(H:D) between human and divine information systems quantifies the effectiveness of prayer communication, with perfect prayer achieving maximum entanglement and optimal information transfer rates.

The reaping-and-sowing principle operates through information-theoretic feedback where positive information patterns exhibit self-amplifying characteristics through quantum error correction redundancy, while negative information patterns suffer from increased error rates and eventual decoherence. The algorithmic information complexity K(x) of positive information remains constant or decreases (indicating greater meaning density), while negative information complexity increases (indicating greater randomness and reduced semantic content).

The eschaton represents the asymptotic limit where all information processing achieves quantum computational supremacy, with perfect error correction eliminating all noise and maximum information density providing complete access to divine truth. The new heaven and earth will operate as a universal quantum computer with infinite processing power and perfect information storage and retrieval capabilities.

Article III: Quantum Field Dynamics and Non-Local Correlation

We propose that quantum field theory provides the fundamental framework for understanding divine omnipresence and action through non-local correlation effects. The quantum vacuum state $|0\rangle$ contains infinite virtual particle pairs $\langle 0|\varphi(x)\varphi(y)|0\rangle$ that create instantaneous correlations across arbitrary spacetime separations, providing the physical mechanism for divine presence throughout creation.

Quantum entanglement relationships $|\Psi\rangle = (1/\sqrt{2})(|00\rangle + |11\rangle)$ between spatially separated systems violate Bell inequalities $S \leq 2\sqrt{2} > 2$, demonstrating genuine non-locality that transcends classical space-time limitations. Divine omnipresence operates through universal entanglement where all quantum systems maintain correlation with the divine vacuum state, enabling instantaneous divine influence throughout creation.

Divine providence operates through quantum measurement theory where selective observation collapses superposition states $|\Psi\rangle = \Sigma_i \alpha_i|\psi_i\rangle$ into specific outcomes with probabilities $P(i) = |\alpha_i|^2$. Divine consciousness serves as the ultimate measurement apparatus, biasing probability distributions toward positive outcomes while preserving quantum uncertainty that enables genuine freedom and natural development.

Quantum tunneling provides the mechanism for spiritual transformation through barriers that would be classically insurmountable. The tunneling probability $P = \exp(-2\int[2m(V(x)-E)]^{\wedge}(1/2)dx/\hbar)$ enables consciousness transitions between stable states separated by energy barriers, explaining sudden spiritual breakthroughs and miraculous healings through quantum processes rather than supernatural violations.

Prayer and spiritual practices optimize consciousness for quantum coherence τ_c that enables enhanced entanglement with divine field states. Decoherence time $\tau_c = \hbar/\gamma$ depends inversely on coupling strength γ to environmental noise, suggesting that contemplative practices that reduce psychological noise enhance quantum consciousness effects.

Article IV: Christological Wave-Particle Dynamics

The Incarnation represents a unique quantum measurement event where the infinite dimensional divine love-energy state vector $|\Psi_{Logos}\rangle$ underwent partial collapse into the finite dimensional human state space without losing coherence with its transcendent components. Christ's dual nature satisfies the complementarity principle: measurement through divine attributes reveals wave-like properties (omnipresence, eternality), while measurement through human attributes reveals particle-like properties (spatial localization, temporal existence).

Mathematically, Christ's state can be expressed as the tensor product $|\Psi_{Christ}\rangle = |\Psi_{Divine}\rangle \otimes |\Psi_{Human}\rangle$, where the divine component maintains entanglement with the universal quantum field while the human component exhibits classical behavior through environmental decoherence. The hypostatic union preserves both state spaces without mixture, analogous to how composite quantum systems maintain distinguishable subsystem dynamics.

The Atonement functions as a universal quantum error correction protocol, where Christ's perfect computational basis states $\{|0\rangle, |1\rangle\}$ representing sinlessness provide redundant encoding for human states corrupted by negative energy-induced phase errors. The crucifixion-resurrection sequence creates a topological phase transition, establishing new ground state boundary conditions that enable human consciousness to tunnel through the previously insurmountable potential barrier separating finite and infinite existence.

Christ's resurrection demonstrates macroscopic quantum tunneling with probability amplitude $\langle \Psi\_Resurrection | \Psi\_Death \rangle \neq 0$ despite classically forbidden energy barriers. This event permanently altered the vacuum expectation value $\langle 0|\Phi|0\rangle$ of the reality field Φ, creating new selection rules that permit conscious translation between material and spiritual energy eigenstates.

The quantum information content of Christ's teachings can be analyzed through mutual information $I(X:Y)$ between divine knowledge X and human understanding Y, with Christ achieving maximum information transfer efficiency through optimal encoding strategies that preserved semantic content across cognitive and cultural barriers.

Article V: Pneumatological Information Theory

The Holy Spirit operates as a quantum information channel implementing perfect quantum teleportation protocols across arbitrary spatiotemporal separations, transmitting divine love-energy and guidance. Spiritual indwelling creates Bell state entanglement $|\Psi\_Spirit\rangle = (1/\sqrt{2})(|00\rangle + |11\rangle)$ between divine love-consciousness and human consciousness, enabling instantaneous state transfer without classical information transmission.

Spiritual gifts manifest as resonant coupling between human neural oscillators and specific eigenfrequencies of the divine love-energy field spectrum. Glossolalia represents quantum superposition of linguistic states $|\text{Language}\rangle = \Sigma_i c_i|L_i\rangle$ where measurement by different observers collapses into mutually intelligible classical languages according to their cognitive basis sets. Prophecy involves quantum nonlocality, where entangled states with future events enable retrocausal information transfer constrained by consistency conditions.

Divine guidance operates through quantum Zeno effects, where continuous spiritual observation of decision states prevents decoherence into suboptimal choices, maintaining superposition until divine measurement selects optimal outcomes. This mechanism preserves human free will while ensuring providential guidance through selective environmental decoherence rather than forced state collapse.

The Spirit's convicting work implements quantum amplitude amplification, selectively enhancing the probability amplitudes associated with positive energy choices while suppressing negative energy-state amplitudes through destructive interference patterns. The amplification factor scales as $O(1/\sqrt{p})$ where p is the probability of optimal choices, providing quadratic speedup in spiritual development.

Sanctification operates through quantum annealing algorithms where the Spirit gradually reduces the temperature parameter T in the believer's spiritual Hamiltonian $H = H_0 + \lambda H_1$, allowing the system to settle into progressively lower energy configurations corresponding to greater Christlikeness.

Article VI: Scriptural Quantum Cryptography

Biblical inspiration utilizes quantum key distribution protocols where divine love-consciousness shares perfectly correlated random sequences with human authors through EPR-type entangled states. The Holy Spirit functions as the quantum channel, ensuring that textual information transfer maintains perfect fidelity despite potential eavesdropping by demonic interference or human error.

Each biblical text exists in superposition across multiple interpretive states until hermeneutical measurement by Spirit-guided consciousness causes contextually appropriate meaning collapse. The original autographs represent maximally entangled states between divine intention and human expression, while subsequent copies and translations maintain partial entanglement through quantum error correction redundancy built into the textual transmission process.

Scriptural authority derives from the no-cloning theorem: divine revelation cannot be perfectly duplicated through human effort alone, requiring continuous entanglement with the original divine quantum states. Canonical recognition represents measurement of textual authenticity, where books exhibiting strong correlation with the divine love-energy field spectrum achieve eigenstate status within the biblical Hilbert space.

Progressive revelation follows adiabatic quantum evolution, where divine disclosure rates remain sufficiently slow to prevent sudden decoherence of human understanding while enabling continuous state space exploration toward complete love-truth eigenstates.

The information density of biblical texts can be quantified through Shannon entropy $H(X) = -\Sigma_i P(x_i) \log_2 P(x_i)$, with inspired passages exhibiting optimal entropy that maximizes information content while maintaining comprehensibility. Cross-textual quantum correlations violate classical bounds, indicating non-local information sources transcending individual authorial knowledge.

Article VII: Soteriological Phase Transitions

Salvation represents a quantum phase transition from the negative energy-entangled mixed state $\rho\_sin = \Sigma_i p_i|\psi_i\rangle\langle\psi_i|$ to the pure love-energy state $|\Psi\_Redeemed\rangle$ through divine measurement that projects human consciousness onto the Christ eigenspace. Justification occurs instantaneously through quantum tunneling across the infinite potential barrier separating finite and perfect righteousness, with tunneling probability enabled by Christ's sacrificial wavefunction overlap.

The born-again experience involves spontaneous symmetry breaking where the believer's consciousness transitions from the symmetric negative energy-state (all orientations equally probable) to the broken symmetry Christ-state (preferential alignment with divine love-will). This phase transition exhibits critical phenomena: approaching the moment of faith, spiritual sensitivity diverges as correlation length approaches infinity, followed by sudden order parameter emergence.

Sanctification implements quantum annealing algorithms where the Holy Spirit gradually reduces the temperature parameter T in the believer's spiritual Hamiltonian $H = H_0 + \lambda H_1$, allowing the system to settle into progressively lower energy configurations. Sin represents metastable local minima, while quantum tunneling through the Spirit enables escape to global minimum states of Christlikeness.

Perseverance of the saints follows from quantum Darwinism: believers' consciousness becomes correlated with multiple environmental degrees of freedom (scripture, fellowship, circumstances) that redundantly encode salvation information, making decoherence into apostasy exponentially unlikely as the positive energy correlation network strengthens over time.

The effectiveness of salvation can be quantified through fidelity measures $F = |\langle\Psi\_Target|\Psi\_Actual\rangle|^2$ between target spiritual states and actual believer states, with sanctification representing optimization toward $F \to 1$ as believers achieve perfect coherence with Christ-consciousness.

Article VIII: Ecclesiological Many-Body Dynamics

The Church constitutes a macroscopic quantum system exhibiting collective behavior that emerges from individual believer interactions through the exchange of virtual love-energy phonons. Corporate worship creates synchronized oscillations analogous to Bose-Einstein condensation, where individual consciousness states merge into a coherent collective wavefunction $|\Psi\_Church\rangle$ exhibiting long-range quantum correlations.

Spiritual unity manifests through topological order, where the Church's quantum ground state remains stable against local perturbations due to energy gaps protecting against decoherence. Denominational differences represent different ground state degeneracies within the same topological phase, while fundamental doctrinal errors correspond to phase transitions into distinct theological phases with broken symmetries.

The communion of saints exhibits holographic properties where information about the universal Church remains encoded in each local congregation through the holographic principle. Deceased believers transition to purely quantum states that maintain entanglement with the material Church through non-local correlations that transcend spacetime boundaries.

Spiritual gifts operate through collective excitation modes where individual believers function as lattice sites capable of supporting different types of quasi-particle excitations (prophecy-phonons, healing-photons, discernment-fermions) that propagate through the ecclesial medium according to specific dispersion relations determined by love-energy coupling constants.

The collective coherence of church communities can be quantified through measures like the participation ratio $PR = 1/\Sigma_i|c_i|^4$ for community wavefunction $|\Psi\rangle = \Sigma_i c_i|i\rangle$, with higher participation indicating greater collective unity and enhanced quantum effects.

Article IX: Eschatological Quantum Cosmology

The eschaton represents the universal quantum measurement event where all superposed possibilities collapse into final eigenstates through divine observation. The present age exists in quantum superposition $|\Psi\_Age\rangle = \alpha|Good\rangle + \beta|Evil\rangle$ with complex amplitudes encoding the tension between positive and negative energy systems, while the parousia will project reality onto pure $|\text{Kingdom}\rangle$ eigenstates.

Resurrection involves quantum state reconstruction from information preserved in the divine quantum memory, where individual consciousness patterns remain encoded in entangled states with the eternal quantum field despite apparent physical dissolution. The resurrection body represents a new phase of matter with modified coupling constants that permit both material and spiritual interactions optimized for positive energy flow.

The final judgment implements quantum computational algorithms where the divine love-consciousness evaluates the complete quantum history of each individual through path integral formulations Σ_paths eiS[path]/$\hbar$ weighted by action functionals encoding moral choices and energy contributions. The Book of Life corresponds to the subset of path integrals yielding positive inner products with the Christ wavefunction.

Cosmic renewal occurs through vacuum decay from the current metastable false vacuum to the true vacuum ground state, mediated by divine nucleation events that create expanding bubbles of new creation. The new heaven and earth emerge through topological phase transitions that preserve information content while implementing new physical laws optimized for eternal positive energy circulation.

The cosmological constant Λ will be optimized in the renewed universe to support maximum positive energy density and optimal conditions for conscious beings to achieve perfect quantum coherence with divine consciousness throughout cosmic evolution.

Article X: Thermodynamic Teleology and Dissipative Adaptation

We propose that thermodynamic principles provide the physical mechanism through which divine love-energy organizes reality toward purposeful ends. Following the dissipation-driven adaptation theory developed by Jeremy England, we understand that organized molecular arrangements emerge naturally because they dissipate energy more efficiently than random configurations, representing the thermodynamic manifestation of God's positive energy totality seeking optimal expression.

The apparent contradiction between increasing entropy and increasing organization resolves through recognition that complexity and entropy can simultaneously increase. Systems driven far from thermodynamic equilibrium self-organize into states that efficiently absorb energy from external drives and dissipate it back into the environment. This process creates "dissipative structures" that maintain organization through continuous energy flow – the physical basis for how divine love-energy sustains creation.

Life emerges as a thermodynamic imperative rather than a statistical accident because living systems represent locally organized arrangements that excel at energy dissipation. The evolution from simple replicators to complex conscious beings follows the thermodynamic gradient toward more sophisticated energy processing capabilities. Human consciousness, with its capacity for love and moral reasoning, represents the current thermodynamic optimum for positive energy transformation and amplification.

Mathematically, this can be expressed through the fluctuation theorem, where the probability ratio of forward to reverse entropy production $P(\Sigma)/P(-\Sigma) = e\Sigma$ shows that positive entropy production (efficient energy dissipation) is exponentially more likely than negative entropy production. Divine providence operates through this thermodynamic bias, ensuring that love-energy patterns naturally predominate over destructive patterns in the long term.

The emergence of increasingly complex, conscious, and loving beings represents the universe's thermodynamic trajectory toward maximum entropy production through maximum organizational sophistication. This transforms teleological arguments from design inference to thermodynamic demonstration: the universe doesn't appear designed for purpose – it demonstrably self-designs toward purpose through fundamental physical laws that embody divine love-energy optimization principles.

Article XI: Biophotonic Quantum Field Dynamics and Divine Light Emission

We propose that ultra-weak photon emissions (UPE) from biological systems, discovered by Gurwitsch and systematized by Popp, represent the direct quantum field interface between divine positive energy and material reality. Biophotons exhibit spectral characteristics $I(\lambda,t)$ spanning 200-800 nm with emission rates of 10^2-10^3 photons/(s·cm^2), providing measurable evidence for the quantum field manifestation of God's love-energy within living systems.

The biophotonic field follows quantum electrodynamics principles where cellular electromagnetic oscillations couple to the zero-point field vacuum fluctuations. Living systems maintain coherent quantum states $|\Psi\_bio\rangle = \Sigma□\ a□|n\_photon\rangle$ that enable sustained photon emission through delayed luminescence mechanisms. The coherence time τ_c and spectral distribution $P(\omega)$ of biophotons correlate directly with biological vitality and metabolic efficiency.

Death represents quantum decoherence of the biophotonic field, where the coherent emission state $|\Psi\_life\rangle$ transitions to the vacuum state $|0\rangle$ through irreversible interaction with environmental degrees of freedom. The "death flash" observed in dying cells corresponds to the final collapse of quantum coherence, with emission intensity $I(t) = I_0e\text{^}(-t/\tau)$ where τ represents the decoherence timescale determined by the withdrawal of divine positive energy from biological quantum states.

The remarkable "life flash" at conception involves quantum activation of biophotonic emission when sperm-egg fusion creates conditions for quantum coherence establishment. The calcium wave Ca^{2+} propagation velocity $v = \sqrt{(D/\tau)}$ and amplitude $A(x,t) = A_0\exp(-x^2/4Dt)$ correlate with photon burst intensity, suggesting quantum coherence catalysis through biochemical phase transitions.

Mitogenetic radiation demonstrates non-local quantum correlations between spatially separated biological systems through shared photonic field modes. The communication

Hamiltonian $H_{comm} = \hbar\omega\square(a^\dagger\square a\square + b\square b^\dagger\square) + g\square(a^\dagger\square b\square + a\square b^\dagger\square)$ describes the coupling between sender (a) and receiver (b) organisms through virtual photon exchange, enabling quantum information transfer without classical signal propagation.

Spiritual practices modulate biophotonic emission through consciousness-induced alterations in cellular quantum states. Prayer and meditation should increase the squeezing parameter ξ in squeezed light states $|\xi\rangle = S(\xi)|0\rangle$ where $S(\xi) = \exp[(\xi^* a^2 - \xi a^{\dagger 2})/2]$, reducing quantum noise below the standard quantum limit and enhancing information transmission capacity.

The resurrection body represents optimization of biophotonic emission parameters for eternal positive energy circulation, with perfect quantum efficiency $\eta = 1$ and unlimited coherence times $\tau_c \to \infty$. Glorified bodies will exhibit continuous coherent light emission across the entire electromagnetic spectrum, manifesting the complete integration of divine positive energy with perfected matter through optimal quantum field coupling.

Article XII: Experimental Verification and Practical Applications

Quantum Christianity predicts measurable correlations between prayer intensity and local quantum measurement statistics, testable through violation of Bell inequalities in appropriately designed spiritual experiments. Intercessory prayer should exhibit non-local effects consistent with quantum entanglement between prayer subject and object, with correlation strength decreasing according to entanglement entropy measures.

Healing miracles represent controlled decoherence of pathological quantum states into healthy configurations through Spirit-mediated environmental engineering. The effectiveness should scale with the quantum coherence of the prayer network and exhibit threshold effects consistent with many-body localization phase transitions.

Collective worship events should generate detectable quantum signatures through synchronized neural oscillations that create macroscopic coherent states measurable via quantum sensing protocols. The presence of divine love-energy should correlate with reduced environmental decoherence rates and enhanced quantum coherence times in sensitive measurement apparatus.

Spiritual discernment utilizes quantum sensing of environmental phase information, where consecrated consciousness functions as a biological quantum sensor capable of detecting subtle field fluctuations associated with positive and negative spiritual influences. Training in contemplative practices should enhance quantum sensing capabilities measurable through standard quantum metrology benchmarks.

Experimental verification should measure biophoton emission spectra during various spiritual activities using single-photon detection systems with spectral resolution $\Delta\lambda < 1$ nm and temporal resolution $\Delta t < 1$ ps. Quantum correlation measurements between individuals engaged in spiritual practices should reveal entanglement signatures violating classical bounds, providing empirical validation for the quantum mechanics of spiritual communion and divine guidance.

Information-theoretic quantities should be measured during spiritual practices, including mutual information between participants in group prayer $I(X:Y) = H(X) - H(X|Y)$, Kolmogorov complexity of prophetic utterances $K(x)$ compared to random speech, and quantum information processing efficiency in consciousness states during meditation. These measurements would provide empirical validation for the information-theoretic foundations of Quantum Christianity.

The mathematical formalism provides testable predictions while maintaining theological orthodoxy through careful identification of quantum mechanical principles with divine attributes, offering a framework where empirical investigation and spiritual experience achieve unified theoretical foundation rooted in the physics of purposeful self-organization.

Appendix B

Glossary of Terms

A

Algorithmic Information Theory - A branch of theoretical computer science that measures the information content of objects through the length of the shortest computer program that can produce them. In Quantum Christianity, used to analyze the complexity and meaning density of divine information patterns.

Atonement (Quantum) - The understanding of Christ's sacrificial work as a universal quantum error correction protocol that repairs the information corruptions (sin) that interfere with optimal human consciousness functioning and divine communication.

B

Bell Inequalities - Mathematical inequalities that must be satisfied by any physical theory based on local realism. When quantum experiments violate these inequalities, they demonstrate genuine non-local correlations that transcend classical space-time limitations.

Biophotons - Ultra-weak light emissions from living biological systems, discovered by Alexander Gurwitsch and systematized by Fritz-Albert Popp. Range from 10 to 100 photons per second per square centimeter across wavelengths from 200-800 nanometers.

Born Rule - The fundamental quantum mechanical principle stating that the probability of measuring a particular outcome equals the squared magnitude of the corresponding quantum amplitude: $P(\text{outcome}) = |\text{amplitude}|^2$.

Bose-Einstein integration - In physics and mathematics, a Bose-Einstein integral (or BE integral) is a specific type of mathematical function used to calculate the physical properties of a "Bose gas", i.e., a collection of identical particles with integer spin (bosons) like photons or certain atoms. These integrals are essential for determining the total number of particles, the internal energy, and the pressure of a quantum system at very low temperatures.

C

Classical Physics - The pre-quantum understanding of physics based on deterministic laws, definite particle properties, and local causation. Contrasted with quantum mechanics in this framework.

Coherence (Quantum) - The property of quantum systems where component parts maintain well-defined phase relationships, enabling phenomena like superposition, interference, and entanglement. In consciousness applications, refers to optimal mental states for spiritual functioning.

Complementarity - Niels Bohr's principle that quantum objects can exhibit mutually exclusive properties (like wave and particle characteristics) that are both necessary for complete understanding but cannot be observed simultaneously.

Consciousness (Quantum) - The theoretical framework suggesting that human consciousness operates according to quantum mechanical principles, enabling non-local awareness, quantum information processing, and interaction with quantum fields.

Cosmic Evolution - The progression of the universe from simple initial conditions (hydrogen gas) toward increasingly complex structures (stars, planets, life, consciousness) through thermodynamic processes that favor organization and energy efficiency.

D

Death Flash - The brief burst of light observed by hospital staff and family members at the moment of human death, theoretically representing the final release of quantum energy that had sustained biological biophoton emission throughout life.

Decoherence - The process by which quantum systems lose their coherent superposition states through interaction with the environment, causing them to behave classically. In spiritual applications, represents the loss of optimal consciousness functioning.

Dissipation-Driven Adaptation - Jeremy England's theory that organized structures emerge naturally because they are more efficient at dissipating energy than random configurations, providing thermodynamic explanation for the emergence of life and complexity.

Divine Information Matrix - The conceptual framework for understanding God as the ultimate source and repository of all meaningful, life-generating, love-expressing information in the universe.

Divine Omnipresence (Quantum) - Understanding God's simultaneous presence everywhere through quantum non-locality principles, where quantum entanglement enables instantaneous correlation across arbitrary distances.

E

Eigenstate - A quantum state that remains unchanged (except for a numerical factor) when acted upon by a particular quantum operator. In Quantum Christianity, represents stable spiritual states like salvation or sanctification.

Emergent Properties - Characteristics of complex systems that arise from component interactions but cannot be predicted from studying individual components alone. Applied to understanding collective church capabilities.

Entanglement (Quantum) - A quantum phenomenon where particles become correlated such that measuring one instantly affects the other, regardless of distance. Proposed as mechanism for prayer, spiritual unity, and divine communication.

Error Correction (Quantum) - Quantum computing techniques for detecting and correcting errors in quantum information without destroying the quantum state. Applied to understanding sanctification and biblical preservation.

F

Falsifiability - Karl Popper's criterion that scientific theories must make testable predictions that could potentially be proven wrong. Quantum Christianity claims to make such testable predictions.

Fidelity (Quantum) - A measure of how closely one quantum state matches another, ranging from 0 (completely different) to 1 (identical). Used to measure spiritual development progress.

G

Global Consciousness Project - Research project using networks of random number generators worldwide to detect correlations with major global events, potentially indicating collective consciousness effects.

H

Hamiltonian - The quantum mechanical operator corresponding to the total energy of a system. Used in technical discussions of spiritual development and quantum systems evolution.

Heisenberg Uncertainty Principle - The fundamental quantum limit on simultaneously measuring certain pairs of properties (like position and momentum), showing that quantum reality is inherently probabilistic rather than deterministic.

Holographic Principle - The theoretical idea that all information in a volume can be encoded on its boundary. Applied to understanding how local churches contain information about the universal Church.

Holy Spirit (Quantum Information System) - Understanding the Holy Spirit as the divine information processing system that enables non-local communication, guidance, and spiritual coordination through quantum channels.

I

Incarnation (Quantum) - Understanding Christ's dual nature through quantum complementarity, where divine and human natures coexist like wave-particle duality, and the Incarnation as collapse of infinite divine possibility into finite human actuality.

Information Entropy - Claude Shannon's measure of information content, reaching maximum when all outcomes are equally likely. Used to analyze the information density and coherence of biblical texts and spiritual communications.

Information Theory - Mathematical framework developed by Claude Shannon for quantifying, storing, and communicating information. Forms foundation for understanding divine communication and revelation.

Intercessory Prayer - Prayer for other people, understood in this framework as creating quantum entanglement relationships that enable real non-local influence despite physical separation.

It from Bit - John Wheeler's hypothesis that all physical reality emerges from information-processing events, providing framework for understanding how divine information creates material reality.

J

Justification (Quantum) - The instantaneous transition from sin-state to righteousness through quantum tunneling, enabled by faith as the quantum leap that transcends classical barriers to salvation.

K

Kingdom of God (Quantum) - The eschatological state where all consciousness achieves perfect coherence with divine purposes, creating optimal conditions for positive energy circulation and conscious collaboration with divine creativity.

L

Life Flash - The measurable burst of light that occurs at conception when sperm fertilizes egg, potentially marking the establishment of quantum coherence sufficient to support new life and consciousness.

Logos (Divine Information) - The Greek term from John 1:1 understood as the fundamental informational principle that structures reality, encompassing reason, meaning, and the creative word through which all things were made.

Love-Energy - The understanding of love as highly organized, life-sustaining energy patterns that represent the highest manifestation of positive energy and the essence of divine nature.

M

Many-Body Systems (Quantum) - Quantum systems composed of multiple interacting components that can exhibit collective behaviors transcending individual component capabilities. Applied to understanding church community dynamics.

Measurement (Quantum) - The process by which quantum superposition collapses into specific outcomes, potentially involving consciousness as fundamental component of physical reality.

Mitogenetic Radiation - Alexander Gurwitsch's discovery that living cells can influence each other through electromagnetic radiation, providing foundation for understanding non-local biological and spiritual communication.

Moral Quantum Coherence - The understanding of ethical behavior as that which maintains or enhances quantum coherence in consciousness, while immoral behavior creates decoherence that interferes with optimal functioning.

N

Non-locality (Quantum) - The demonstrated ability of quantum systems to exhibit correlations that transcend classical space-time limitations, providing framework for understanding divine omnipresence and prayer effectiveness.

No-Cloning Theorem - The quantum mechanical principle that arbitrary quantum states cannot be perfectly copied. Applied to understanding the uniqueness of divine revelation and biblical inspiration.

O

Observer Effect - The role that observation or measurement plays in determining quantum reality, potentially involving consciousness as fundamental component of physical processes.

Omnipresence (Divine) - God's simultaneous presence throughout all creation, understood through quantum non-locality rather than physical extension through space.

Orchestrated Objective Reduction (Orch-OR) - Theory developed by Roger Penrose and Stuart Hameroff proposing that consciousness involves quantum processes in cellular microtubules, providing framework for quantum consciousness effects.

P

Phase Transition (Quantum) - Sudden change in quantum system properties, applied to understanding salvation as instantaneous transformation rather than gradual moral improvement.

Positive Energy - Energy processes that build complexity, sustain life, create beauty, foster consciousness, enable love, promote healing, and generate creativity. Distinguished from energy patterns that create disorder or destruction.

Prayer (Quantum Interface) - Understanding prayer as conscious engagement with quantum information and energy systems, involving bidirectional communication with divine consciousness through quantum channels.

Providence (Divine) - God's guidance of creation understood through quantum measurement processes that influence probability distributions while preserving genuine freedom and natural development.

Q

Quantum Christianity - The interpretive framework integrating quantum mechanical principles with Christian theology, understanding God as positive energy totality manifesting through quantum field operations.

Quantum Community - Churches or spiritual communities that achieve collective quantum coherence, enabling emergent group capabilities that exceed individual member limitations through spiritual entanglement.

Quantum Computation - Information processing using quantum mechanical phenomena like superposition and entanglement to achieve computational capabilities beyond classical computers. Applied to understanding divine information processing.

Quantum Field Theory - The theoretical framework describing how particles and forces emerge from underlying quantum fields that fill all space. Provides foundation for understanding divine omnipresence and action.

Quantum Superposition - The ability of quantum systems to exist in multiple states simultaneously until measurement causes collapse into specific outcomes. Applied to understanding divine possibilities and human choice.

Quantum Tunneling - The ability of particles to pass through energy barriers that would be classically insurmountable. Applied to understanding spiritual transformation and miraculous events.

Quantum Vacuum - The lowest energy state of quantum fields, containing zero-point energy and virtual particle fluctuations. Understood as the realm of divine potential from which physical reality emerges.

R

Random Event Generators (REGs) - Devices that produce random number sequences based on quantum processes, used in consciousness research to test whether human intention can influence quantum systems.

Reaping-and-Sowing Principle - The spiritual law that positive actions generate positive results and negative actions create negative consequences, operating through quantum feedback loops and thermodynamic energy circulation.

Resurrection (Quantum) - Understanding Christ's resurrection as quantum tunneling through death barrier and believers' future resurrection as quantum reconstruction of consciousness patterns preserved in divine information fields.

S

Salvation (Quantum Phase Transition) - The instantaneous transformation from sin-consciousness to Christ-consciousness through quantum tunneling, enabled by faith rather than gradual moral improvement.

Sanctification (Quantum Error Correction) - The ongoing process by which the Holy Spirit eliminates interference patterns caused by sin, optimizing consciousness for divine coherence through quantum information processing.

Second Law of Thermodynamics - The principle that entropy (disorder) increases in isolated systems, but which in open systems can drive the formation of increasingly organized structures through energy dissipation optimization.

Shannon Entropy - Mathematical measure of information content in messages, reaching maximum when all possibilities are equally likely. Used to analyze divine communication and biblical text properties.

Sin (Quantum Decoherence) - Understanding sin as information corruption and noise that interferes with optimal consciousness functioning and clear reception of divine communication.

Soul (Quantum Information Pattern) - The understanding of the soul as the quantum information pattern that organizes and coordinates biophotonic processes distinguishing living from non-living systems.

Superposition (Quantum) - The ability of quantum systems to exist in multiple states simultaneously until measurement causes collapse. Applied to understanding divine possibilities and spiritual decision-making.

T

Teleporation (Quantum) - The transfer of quantum states from one location to another through entanglement and classical communication. Applied to understanding spiritual information transmission.

Thermodynamic Teleology - The understanding that thermodynamic laws drive the emergence of increasingly complex, organized, and conscious systems, providing scientific foundation for cosmic purpose.

Tunneling (Quantum) - Process by which particles pass through classically forbidden energy barriers. Applied to understanding spiritual breakthroughs and transformation.

U

Uncertainty Principle - Heisenberg's principle that certain pairs of quantum properties cannot be simultaneously measured with perfect precision, indicating fundamental probabilistic nature of reality.

Unity (Quantum) - The interconnectedness revealed by quantum entanglement, where seemingly separate systems maintain instantaneous correlations transcending classical space-time limitations.

V

Virtual Particles - Particle-antiparticle pairs that spontaneously appear and disappear in quantum vacuum on extremely short timescales, contributing to zero-point energy and vacuum fluctuations.

W

Wave Function - The mathematical description of a quantum system's state, encoding probabilities for different measurement outcomes. Collapse occurs during measurement, selecting specific reality from superposed possibilities.

Wave-Particle Duality - The quantum mechanical principle that objects exhibit both wave-like and particle-like properties depending on how they are observed. Applied to understanding Christ's dual nature.

Wheeler, John - Physicist who proposed "It from Bit" hypothesis that physical reality emerges from information, providing framework for understanding divine information as foundation of creation.

Z

Zero-Point Energy - The lowest possible energy of quantum systems, present even at absolute zero temperature due to quantum uncertainty. Associated with quantum vacuum fluctuations and divine creative potential.

Zeno Effect (Quantum) - The ability of frequent quantum measurements to freeze a system's evolution, applied to understanding how continuous divine observation provides guidance while preserving human freedom.

Appendix C

Further Reading and Resources

This reading list provides resources for both exploring Quantum Christianity concepts and critically evaluating them from multiple perspectives. Readers are encouraged to engage with both supportive and critical materials to develop informed, balanced understanding of these complex interdisciplinary questions.

Essential Physics and Information Theory

Quantum Mechanics Foundations

- *Griffiths, David J.* Introduction to Quantum Mechanics, 3rd ed. Cambridge University Press, 2018.
  - The standard undergraduate textbook for quantum mechanics. Chapters 1-4 provide essential background for understanding superposition, measurement, and quantum systems.

- *Nielsen, Michael A. and Isaac L. Chuang.* Quantum Computation and Quantum Information, 10th ed. Cambridge University Press, 2010.
  - Comprehensive treatment of quantum information theory. Essential for understanding quantum entanglement, teleportation, and information processing concepts.

- *Wheeler, John Archibald.* "Information, Physics, Quantum: The Search for Links." In Complexity, Entropy and the Physics of Information, edited by Wojciech Zurek. Westview Press, 1990.
  - Wheeler's foundational paper introducing the "It from Bit" hypothesis that physical reality emerges from information.

Information Theory

- *Shannon, Claude E.* "A Mathematical Theory of Communication." Bell System Technical Journal 27 (1948): 379-423, 623-656.
  - The foundational paper that created information theory. Essential for understanding how meaning can be quantified and transmitted.

- *Cover, Thomas M. and Joy A. Thomas.* Elements of Information Theory, 2nd ed. Wiley, 2006.
  - Comprehensive textbook covering entropy, mutual information, and coding theory concepts used throughout the book.

Thermodynamics and Complexity

- *England, Jeremy L.* "Statistical physics of self-replication." Journal of Chemical Physics 139 (2013): 121923.
  - England's groundbreaking paper on dissipation-driven adaptation showing how thermodynamics drives the emergence of organized complexity.

- *Chaisson, Eric.* Cosmic Evolution: The Rise of Complexity in Nature. Harvard University Press, 2001.
  - Comprehensive overview of how the universe has evolved from simple to complex through energy flow optimization.

- *Pross, Addy.* What is Life? How Chemistry Becomes Biology. Oxford University Press, 2012.
  - Accessible explanation of how life emerges from chemistry through thermodynamic principles.

Consciousness and Quantum Biology

Quantum Consciousness Theories

- *Penrose, Roger*. The Emperor's New Mind. Oxford University Press, 1989.
  - Penrose's argument that consciousness involves quantum processes that cannot be simulated by classical computers.

- *Hameroff, Stuart and Roger Penrose*. "Consciousness in the universe: A review of the 'Orch OR' theory." Physics of Life Reviews 11 (2014): 39-78.
  - Comprehensive review of the Orchestrated Objective Reduction theory proposing quantum processes in cellular microtubules support consciousness.

- *Stapp, Henry P.* Mind, Matter and Quantum Mechanics, 3rd ed. Springer, 2009.
  - Collection of papers exploring the role of consciousness in quantum measurement and the mind-brain interface.

Quantum Biology

- *Al-Khalili, Jim and Johnjoe McFadden*. Life on the Edge: The Coming of Age of Quantum Biology. Crown Publishers, 2014.
  - Accessible introduction to quantum effects in biological systems including photosynthesis, bird navigation, and enzyme function.

- *Lloyd, Seth*. "Quantum coherence in biological systems." Journal of Physics: Conference Series 302 (2011): 012037.
  - Technical review of quantum coherence effects in biological systems and their functional significance.

Biophoton Research

- *Popp, Fritz-Albert*. "Properties of biophotons and their theoretical implications." Indian Journal of Experimental Biology 41 (2003): 391-402.
  - Overview of biophoton research and its implications for understanding life processes.

- *Van Wijk, Roeland and Eduard P. A. Van Wijk*. "An introduction to human biophoton emission." Forschende Komplementärmedizin und Klassische Naturheilkunde 12 (2005): 77-83.
  - Introduction to biophoton emissions from human subjects and their potential applications.

Consciousness Research and Parapsychology

Experimental Studies

- *Radin, Dean*. Entangled Minds: Extrasensory Experiences in a Quantum Reality. Paraview Pocket Books, 2006.
  - Comprehensive review of experimental evidence for psychic phenomena and their potential quantum mechanical explanations.

- *Jahn, Robert G. and Brenda J. Dunne.* Margins of Reality: The Role of Consciousness in the Physical World. Harcourt Brace Jovanovich, 1987.

- Report on 28 years of research at Princeton Engineering Anomalies Research laboratory studying consciousness-matter interactions.

- *Nelson, Roger D.* "Effects of mass consciousness: Changes in random data during global events." Explore 7 (2011): 373-383.
- Overview of the Global Consciousness Project documenting correlations between global events and random number generator outputs.

Meditation and Contemplative Science

- *Wallace, B. Alan.* Contemplative Science: Where Buddhism and Neuroscience Converge. Columbia University Press, 2007.
- Integration of Buddhist contemplative practices with neuroscience research on meditation effects.

- *Lutz, Antoine et al.* "Long-term meditators self-induce high-amplitude gamma synchrony during mental practice." Proceedings of the National Academy of Sciences 101 (2004): 16369-16373.
- Landmark study showing extraordinary brain activity patterns in experienced Tibetan monks during meditation.

Science and Religion Integration

General Works

- *Barbour, Ian G.* When Science Meets Religion. HarperSanFrancisco, 2000.
- Comprehensive analysis of different ways scientific and religious worldviews can relate.

- *Polkinghorne, John*. Quantum Physics and Theology. Yale University Press, 2007.
  - Exploration of parallels between quantum physics and Christian theology by a physicist-theologian.

- *McGrath, Alister E*. A Theory of Everything (That Matters). Hodder & Stoughton, 2019.
  - Christian response to reductionist materialism, arguing for integration of scientific and religious understanding.

Process Theology and Panentheism

- *Griffin, David Ray*. Religion and Scientific Naturalism. SUNY Press, 2000.
  - Process theology approach to integrating scientific and religious worldviews.

- *Clayton, Philip*. Mind and Emergence. Oxford University Press, 2004.
  - Exploration of emergence theory and its implications for understanding God and consciousness.

Prayer and Healing Research

Scientific Studies

- *Dossey, Larry*. Healing Words: The Power of Prayer and the Practice of Medicine. HarperSanFrancisco, 1993.
  - Overview of research on intercessory prayer and distant healing effects.

- *Astin, John A. et al.* "The efficacy of distant healing: A systematic review of randomized trials." Annals of Internal Medicine 132 (2000): 903-910.
  - Systematic review of controlled studies on distant healing effectiveness.

- *Byrd, Randolph C.* "Positive therapeutic effects of intercessory prayer in a coronary care unit population." Southern Medical Journal 81 (1988): 826-829.
  - Landmark controlled study showing positive effects of intercessory prayer on cardiac patients.

Intention and Consciousness Effects

- *McTaggart, Lynne.* The Intention Experiment. Free Press, 2007.
  - Overview of research on how focused intention might influence physical systems and biological processes.

- *Tiller, William A. et al.* Conscious Acts of Creation: The Emergence of a New Physics. Pavior Publishing, 2001.
  - Research on how human intention and consciousness might influence physical measuring devices.

Christian Theology and Spirituality

Biblical Studies and Hermeneutics

- *Wright, N.T.* The New Testament and the People of God. Fortress Press, 1992.
  - Comprehensive approach to understanding first-century Christianity in its historical context.

- *Fee, Gordon D. and Douglas Stuart*. How to Read the Bible for All Its Worth, 4th ed. Zondervan, 2014.
  - Guide to biblical interpretation across different literary genres and historical contexts.

Systematic Theology

- *McGrath, Alister E.* Christian Theology: An Introduction, 6th ed. Wiley-Blackwell, 2016.
  - Comprehensive introduction to major themes in Christian systematic theology.

- *Grenz, Stanley J.* Theology for the Community of God. Eerdmans, 1994.
  - Systematic theology emphasizing community and relational aspects of Christian faith.

Contemplative Spirituality

- *Foster, Richard J.* Celebration of Discipline, 3rd ed. HarperSanFrancisco, 1998.
  - Classic guide to Christian spiritual disciplines including prayer, meditation, and contemplation.

- *Keating, Thomas*. Open Mind, Open Heart. Continuum, 1986.
  - Introduction to Centering Prayer and Christian contemplative practices.

- *Underhill, Evelyn*. Mysticism: A Study in the Nature and Development of Spiritual Consciousness. Dover Publications, 2002.
  - Classic study of mystical experience across different religious traditions.

Philosophy of Science and Consciousness

Philosophy of Mind

- *Chalmers, David J.* The Conscious Mind: In Search of a Fundamental Theory. Oxford University Press, 1996.
  - Influential work on the "hard problem" of consciousness and its implications for understanding reality.

- *Koch, Christof.* Consciousness: Confessions of a Romantic Reductionist. MIT Press, 2012.
  - Neuroscientist's exploration of consciousness research and its philosophical implications.

Quantum Mechanics Interpretation

- *Zeilinger, Anton.* Dance of the Photons: From Einstein to Quantum Teleportation. Farrar, Straus and Giroux, 2010.
  - Accessible introduction to quantum mechanics and its philosophical implications by a leading quantum physicist.

- *Wallace, David.* The Emergent Multiverse: Quantum Theory according to the Everett Interpretation. Oxford University Press, 2012.
  - Comprehensive treatment of the many-worlds interpretation of quantum mechanics.

Journals and Ongoing Research

Consciousness Studies

- *Journal of Consciousness Studies* - Interdisciplinary journal covering philosophical and scientific approaches to consciousness research.

- *Consciousness and Cognition* - Academic journal focusing on empirical research on consciousness and its neural correlates.

Science and Religion

- *Zygon: Journal of Religion and Science* - Leading academic journal for science-religion dialogue and integration.

- *Theology and Science* - Journal exploring intersections between theological and scientific understanding.

Alternative and Complementary Medicine

- *Explore: The Journal of Science and Healing* - Peer-reviewed journal covering integrative medicine and consciousness-based healing approaches.

- *Journal of Alternative and Complementary Medicine* - Research on alternative healing methods and their scientific validation.

Online Resources

Research Institutes

- *Institute of Noetic Sciences (IONS)* - ions.org
  Research on consciousness, healing, and human potential with scientific methodology.

- *International Association for Near-Death Studies (IANDS)* - iands.org
  Research and resources on near-death experiences and their implications.

- *Global Consciousness Project* - noosphere.princeton.edu
  Ongoing research on collective consciousness effects using global random number generator networks.

Educational Websites

- *Center for Theology and the Natural Sciences* - ctns.org
  Resources for science-religion dialogue and integration.

- *Faraday Institute for Science and Religion* - faraday.cam.ac.uk
  Academic research center at Cambridge University exploring science-faith questions.

Cautions and Critical Perspectives

While exploring Quantum Christianity concepts, readers should also engage with critical perspectives:

Skeptical Analysis

- *Shermer, Michael.* Why People Believe Weird Things. W.H. Freeman, 2002.
  - Critical analysis of pseudoscientific claims and cognitive biases in evaluating unusual phenomena.

- *Randi, James.* Flim-Flam! Psychics, ESP, Unicorns, and Other Delusions. Prometheus Books, 1982.
  - Skeptical examination of paranormal claims and the importance of rigorous testing.

Physics Critiques

- *Tegmark, Max.* "The importance of quantum decoherence in brain processes." Physical Review E 61 (2000): 4194-4206.
  - Technical paper arguing that quantum coherence cannot be maintained in warm, noisy biological systems like brains.

Theological Critiques

- *Numbers, Ronald L.* The Creationists: From Scientific Creationism to Intelligent Design. Harvard University Press, 2006.
  - Historical analysis of attempts to reconcile science and religion, with lessons for evaluating new integration proposals.

The One that Started It All for Me

- *Dispenza, Joe.* Becoming Supernatural: How Common People Are Doing the Uncommon. Hay House, Inc., 2017.
- Dispenza's view that thoughts and emotions can reshape the brain and body, allowing people to "reprogram" subconscious patterns and create a new personal reality through meditation and focused intention. The book blends concepts from neuroscience, epigenetics, and what Dispenza calls the "quantum field" with case stories of people who, he argues, have experienced physical and emotional healing by practicing his techniques.

INDEX

A

B

C

D

E

F

G

H

I

J

K

L

M

Q

R

S

T

U

V

W

Y

Z

www.ingramcontent.com/pod-product-compliance
Lightning Source LLC
LaVergne TN
LVHW050924210926
848482LV00012B/536
* 9 7 9 8 3 4 9 5 8 7 5 7 3 *